Chemical and Biological TERRORISM

Research and Development to Improve Civilian Medical Response

Committee on R&D Needs for Improving Civilian Medical
Response to Chemical and Biological Terrorism Incidents

Health Science Policy Program

INSTITUTE OF MEDICINE

and

Board on Environmental Studies and Toxicology

Commission on Life Sciences

NATIONAL RESEARCH COUNCIL

NATIONAL ACADEMY PRESS
Washington, D.C. 1999

NATIONAL ACADEMY PRESS • 2101 Constitution Avenue, NW • Washington, DC 20418

NOTICE: The project that is the subject of this report was approved by the Governing Board of the National Research Council, whose members are drawn from the councils of the National Academy of Sciences, the National Academy of Engineering, and the Institute of Medicine. The members of the committee responsible for this report were chosen for their special competences and with regard for appropriate balance.

The Institute of Medicine was chartered in 1970 by the National Academy of Sciences to enlist distinguished members of the appropriate professions in the examination of policy matters pertaining to the health of the public. In this, the Institute acts under both the Academy's 1863 congressional charter responsibility to be an adviser to the federal government and its own initiative in identifying issues of medical care, research, and education. Dr. Kenneth I. Shine is president of the Institute of Medicine.

The National Research Council was organized by the National Academy of Sciences in 1916 to associate the broad community of science and technology with the Academy's purposes of furthering knowledge and advising the federal government. Functioning in accordance with general policies determined by the Academy, the Council has become the principal operating agency of both the National Academy of Sciences and the National Academy of Engineering in providing services to the government, the public, and the scientific communities. The Council is administered jointly by both Academies and the Institute of Medicine. Dr. Bruce M. Alberts and Dr. William A. Wulf are chairman and vice chairman, respectively, of the National Research Council.

Support for this project was provided by the Office of Emergency Preparedness, Department of Health and Human Services (Contract No. 282-97-0017). This support does not constitute an endorsement of the views expressed in the report.

Library of Congress Cataloging-in-Publication Data

Chemical and biological terrorism : research and development to improve civilian medical response / Committee on R&D Needs for Improving Civilian Medical Response to Chemical and Biological Terrorism Incidents, Health Science Policy Program, Institute of Medicine, and Board on Environmental Studies and Toxicology, Commission on Life Sciences, National Research Council.
    p. cm.
Includes bibliographical references and index.
ISBN 0-309-06195-4 (hardcover)
    1. Chemical warfare—Health aspects. 2. Biological warfare—Health aspects. 3. Civil defense—United States. 4. Terrorism—Government policy—United States. 5. Disaster medicine—United States. I. Institute of Medicine (U.S.). Committee on R & D Needs for Improving Civilian Medical Response to Chemical and Biological Terrorism Incidents. II. National Research Council (U.S.). Board on Environmental Studies and Toxicology.
    RA648 .C525 1999
    358'.3—dc21

                 98-58069

Additional copies of this report are available for sale from the National Academy Press, 2101 Constitution Avenue, N.W., Box 285, Washington, DC 20055. Call (800) 624-6242 or (202) 334-3313 (in the Washington metropolitan area).

This report is also available online at **http://www.nap.edu.**

For more information about the Institute of Medicine, visit the IOM home page at **http://www2.nas.edu/iom.**

COMMITTEE ON R&D NEEDS FOR IMPROVING CIVILIAN MEDICAL RESPONSE TO CHEMICAL AND BIOLOGICAL TERRORISM INCIDENTS

PETER ROSEN (*Chair*), Director, Emergency Medicine Residency Program, School of Medicine, University of California, San Diego

LEO G. ABOOD, Professor of Pharmacology, Department of Pharmacology and Physiology, University of Rochester Medical Center*

GEORGES C. BENJAMIN, Deputy Secretary, Public Health Services, Department of Health and Mental Hygiene, Baltimore, Maryland

ROSEMARIE BOWLER, Assistant Professor and Fieldwork Coordinator, Department of Psychology, San Francisco State University

JEFFREY I. DANIELS, Leader, Risk Sciences Group, Health and Ecological Assessment Division, Earth and Environmental Sciences Directorate, Lawrence Livermore National Laboratory, Livermore, California

CRAIG A. DeATLEY, Associate Professor, Department of Emergency Medicine and Health Care Sciences Program, The George Washington University, Washington, D.C.

LEWIS R. GOLDFRANK, Director, Emergency Medicine, New York University School of Medicine and Bellevue Hospital Center, New York

JEROME M. HAUER, Director, Office of Emergency Management, City of New York

KAREN I. LARSON, Toxicologist, Office of Toxic Substances, Washington Department of Health, Olympia

MATTHEW S. MESELSON, Thomas Dudley Cabot Professor of the Natural Sciences, Department of Molecular and Cellular Biology, Harvard University, Cambridge, Massachusetts

DAVID H. MOORE, Director, Medical Toxicology Programs, Battelle Edgewood Operations, Bel Air, Maryland

DENNIS M. PERROTTA, Chief, Bureau of Epidemiology, Texas Department of Health, Austin

LINDA S. POWERS, Professor of Electrical and Biological Engineering, and Director, National Center for the Design of Molecular Function, Utah State University, Logan

PHILIP K. RUSSELL, Professor of International Health, School of Hygiene and Public Health, Johns Hopkins University, Baltimore, Maryland

JEROME S. SCHULTZ, Director, Center for Biotechnology and Bioengineering, University of Pittsburgh

ROBERT E. SHOPE, Professor of Pathology, University of Texas Medical Branch, Galveston

ROBERT S. THARRATT, Associate Professor of Medicine and Chief, Section of Clinical Pharmacology and Medical Toxicology, Division of Pulmonary and Critical Care Medicine, University of California, Davis Medical Center, Sacramento

---

*Deceased, January 1998.

Independent Report Reviewers

This report has been reviewed in draft form by individuals chosen for their diverse perspectives and technical expertise, in accordance with procedures approved by the NRC's Report Review Committee. The purpose of this independent review is to provide candid and critical comments that will assist the institution in making its published report as sound as possible and to ensure that the report meets institutional standards for objectivity, evidence, and responsiveness to the study charge. The content of the review comments and draft manuscript remain confidential to protect the integrity of the deliberative process. We wish to thank the following individuals for their participation in the review of this report:

JOHN D. BALDESCHWIELER, Professor of Chemistry, California Institute of Technology, Pasadena

DONALD A. HENDERSON, University Distinguished Professor, School of Hygiene and Public Health, Johns Hopkins University, Baltimore, Maryland

DAVID L. HUXSOLL, Dean, School of Veterinary Medicine, Louisiana State University, Baton Rouge

JOSHUA LEDERBERG, Sackler Foundation Scholar, Rockefeller University, New York

H. RICHARD NESSON, Senior Consultant, Partners Health Care System, Inc., Boston

MICHAEL OSTERHOLM, Chief, Acute Disease Epidemiology Section, Minnesota Department of Health, Minneapolis

ANNETTA P. WATSON, Research Staff, Health and Safety Research Division, Oak Ridge National Laboratory, Oak Ridge, Tennessee
MELVIN H. WORTH, Clinical Professor, State University of New York-Brooklyn and Uniformed Services University of Health Sciences, and Institute of Medicine Scholar-in-Residence

The committee would also like to thank the following individuals for their technical reviews of single chapters of the draft report:

ROBERT E. BOYLE, Formerly Technical Advisor, Chemical Warfare and NBC Defense Division, Office of the Deputy Chief of Staff for Operations, Plans, and Policy, Department of the Army, Washington, D.C.
GREGORY G. NOLL, Hildebrand and Noll Associates, Inc., Lancaster, Pennsylvania
ROBERT S. PYNOOS, Professor and Director, Trauma Psychiatry Service, Department of Psychiatry and Biobehavioral Sciences, University of California, Los Angeles
JOSEPH J. VERVIER, Senior Staff Scientist, ENSCO, Inc., Melbourne, Florida, and formerly Technical Director, Edgewood Research , Development and Engineering Center, Aberdeen Proving Ground, Maryland

Although the individuals listed above have provided many constructive comments and suggestions, it must be emphasized that responsibility for the final content of this report rests solely with the authoring committee and the institution.

Preface

American military forces have been struggling with the issue of chemical and biological warfare for decades—a 1917 National Research Council Committee laid the groundwork for the Army Chemical Warfare Service—but it was the attack of the Tokyo subway with the nerve gas sarin in March 1995 that suddenly put the spotlight on the danger to civilians from chemical and biological attacks. The Federal Emergency Management Agency (FEMA) and the Department of Health and Human Services' Office of Emergency Preparedness (OEP), which is responsible for medical services, have an admirable record of helping state and local governments cope with floods, storms, and other disasters, including terrorism, but, fortunately, no direct experience with the consequences of chemical or biological terrorism. In May 1997, the Institute of Medicine was asked to help OEP prepare for the possibility of chemical or biological terrorism, and, with help from the National Research Council's Board on Environmental Studies and Toxicology, formed this committee to provide recommendations for priority research and development (R&D).

In the ensuing year and a half, the committee met four times, heard presentations on existing technology and ongoing R&D, attempted to absorb a virtual mountain of information, and formulated their recommendations. In the process, a number of things became clear to me. I suspect the rest of the committee would agree, but I will exercise the chair's prerogative at this point, and share the view from my perspective.

First, there is no way to prepare in an optimal fashion for a terror incident. There is too low an incidence to justify the enormous financial outlay it would take to optimally prepare every community for every possible incident. Furthermore, there are not enough incidents for any community to acquire enough experience to make a significant impact on response to the next episode.

Second, although there is a sophisticated technology, described within the body of the report, for in-line detection of an opposing forces chemical agent, it will not be possible to select the sites to protect in a civilian setting with such technology, even if the expense could be borne. At best, it might be possible to selectively protect a public arena where the President was to give an address.

Third, there is no guarantee that the terrorist will announce the attack. Without such an announcement, there will be no recognition that a biological attack is occurring until enough cases, including a number of fatalities, are observed and reported to allow recognition of an epidemic of an unusual disease. Since exposed victims will almost certainly not seek medical care in the same facility, the problem becomes compounded even more greatly.*

Fourth, virtually all the militarily important biologic agents present with early clinical symptoms that resemble viral flu syndromes. Since these are the most common form of acute illness, and since they are usually mild and nonserious, it is probable that the early victims of the attack will be unrecognized, and sent home from a physician's office or Emergency Department as a mild viral syndrome. Therefore, in any response planning, it has to be acknowledged that it will be impossible to prevent ALL mortality, no matter how good a technology can be developed, and no matter how much money we are willing to spend to enhance our response.

Fifth, there is a huge gap between detection technology and therapy. There are many biologic agents, and certainly many chemical agents for which there are no known treatments. We should not expect that terrorists will choose the agents for which we are prepared, and for which we have effective treatment, even if they are the easiest to create and disperse, such as anthrax or sarin.

Sixth, the approach that the committee found most useful to consider in making its recommendations was considering how to superimpose a re-

---

*For example, in Wyoming this year (Summer 1998), there has been an epidemic of *E. coli* diarrhea from a contaminated spring that fed the water supply of the small town of Alpine. There were well over a hundred cases that involved 12 states since the tourists who acquired the disease were from many different locations. It took at least two months to find the source of the contamination, and the only reason that the epidemic was recognized as early as it was, is that there were only a *small* number of medical facilities available to the victims.

sponse to a terror attack upon the systems that are already in place to deal with nonterror events. For example, an earthquake, or a chemical spill, or a flu epidemic will all stress and often overload existing medical facilities. There must be systems in place to deal with these problems, not only on a local basis, but when help must be brought in from outside the afflicted area. These are the systems that will be most appropriate to build on for an effective response to an incident of chemical or biological terrorism.

Seventh, communication between the medical community and agencies that gather and analyze intelligence about potential terrorists and attacks is critical. As alluded to above, it will not only shorten the identification issues and lead to more effective responses, but will clearly lower mortality.

There are a number of areas that will not be covered in this report. For example, it was not possible for the committee to discuss every conceivable biological and chemical weapon that might be used in an attack. It is probable that to prepare only for the list of known weapons and most likely agents will take a commitment and a financial expenditure that will exceed the resources of virtually all communities.

The committee's charge did not include making recommendations on organization and training of individuals and groups faced with managing the consequences of a chemical or biological incident, nor on how to equip such persons or groups, nor on what therapeutic options they should choose. Nevertheless, as noted in our interim report, the committee believes that it would be irresponsible to focus solely on R&D while ignoring potentially simpler, faster, or less expensive mechanisms, such as organization, staff, training, and procurement. Examples from our interim report include:

• Survey major metropolitan hospitals on supplies of antidotes, drugs, ventilators, personal protective equipment, decontamination capacity, mass-casualty planning and training, isolation rooms for infectious disease, and familiarity of staff with the effects and treatment of chemical and biological weapons.

• Encourage the CDC to share with the states its database on the location and owners of dangerous biological materials. State health departments in turn should be encouraged, perhaps by education or training on the effects of the agents and medical responses required, to add infections by these materials to their lists of reportable diseases.

• Convene discussions with FDA on the use of investigational products in mass-casualty situations and on acceptable proof of efficacy for products where clinical trials are not ethical or are otherwise impossible.

• Develop incentives for hospitals to be ambulance-receiving hospitals, to stockpile nerve-agent antidotes and selected antitoxins and put them in the hands of first responders (this may require changes to existing laws or regulations in some states), to purchase appropriate personal protective equipment and expandable decontamination facilities and train emergency department personnel in their use.

• Supplement existing state and federal training initiatives with a program to incorporate existing information on possible chemical or biological terror agents and their treatment into the manuals, SOPs, and reference libraries of first responders, emergency departments, and poison control centers. Professional societies and journal publishers should be recruited to help in this effort.

• Intensify Public Health Service efforts to organize and equip Metropolitan Medical Strike Teams in high-risk cities throughout the country. Although MMSTs are designed to cope with terrorism, because they use local personnel and resources, they also increase the community's general ability to cope with industrial accidents and other mass-casualty events.

Even though the tasks of being prepared and responding adequately appear at times to contain insurmountable obstacles, the committee does believe that by utilizing the resources that are present, along with improvements in communications, monitoring capabilities, detection, and therapeutics, it will be possible to minimize the damage that a terror attack will cause. It is not our intent to leave the readers of this report with feelings of hopelessness. Even if preparation for certain attacks only forces the attackers to choose a weapon that we have not prepared for, we will have developed a system with which we can improvise. The goal, as always in medicine, is to reduce morbidity and mortality and minimize suffering.

In closing I would like to offer my sincere thanks to the staff of the Institute of Medicine, who made our meetings as comfortable and efficient as possible and pulled our sometimes splintered efforts into a coherent whole, and to the members of the committee, busy professionals who volunteered precious time and energy in a highly collegial manner. It was a privilege to work with this outstanding group.

Peter Rosen, M.D.
Chair

Abbreviations

AChE	Acetylcholinesterase
AEL	Acceptable exposure limit
AIDS	Acquired immune deficiency syndrome
APA	American Psychological Association or American Psychiatric Association
ANL	Argonne National Laboratory
ASTM	American Society for Testing and Materials
ATP	adenosine 5′-triphosphate
ATSDR	Agency for Toxic Substances and Disease Registry
BAL	British antiLewisite
BChE	Butyrylcholinesterase
BDO	Battle Dress Overgarment
BIDS	Biological Integrated Detection System
BW	Biological warfare or biological weapon
CAHBS	Civilian Adult Hood Blower System
CAM	Chemical agent monitor
CAPS	Civilian Adult Protective System
CBDCOM	Chemical Biological Defense Command
CBIRF	Chemical Biological Incident Response Force
CBMS	Chemical Biological Mass Spectrometer
CBNP	Chemical and Biological Nonproliferation Program
CBPSS	Chemical Biological Protective Shelter System

C/B-RRT	DoD Chemical/Biological Rapid Response Team
CBWCA	Chemical and Biological Weapons Control Act
CCHF	Crimean Congo hemorrhagic fever
CCP	Crisis Counseling Assistance and Training Program
CDC	Centers for Disease Control and Prevention
cDNA	Complementary (or copy) deoxyribonucleic acid
ChE	Cholinesterase
CISD	Critical incident stress debriefing
CLS	Commission on Life Sciences
CMHS	Center for Mental Health Services
CN⁻	Cyanide anion
CNS	Central nervous system
CSEPP	Chemical Stockpile Emergency Preparedness Program
CSTE	Council of State and Territorial Epidemiologists
CW	Chemical warfare or chemical weapon
CWA	Chemical warfare agent
4-DMAP	4-Dimethylaminophenol
DARPA	Defense Advanced Research Projects Agency
DHHS	Department of Health and Human Services
DMAT	Disaster Medical Assistance Team
DNA	Deoxyribonucleic acid
DNTB	5,5′-dithio-bis (2-nitrobenzoic acid)
DoD	Department of Defense
DoE	Department of Energy
DRN	Disaster Response Network
dsRNA	Double-stranded ribonucleic acid
DSWA	Defense Special Weapons Agency
EDTA	Ethylene diamine tetraacetic acid (dicobalt)
EEE	Eastern equine encephalomyelitis
EF	Edema factor
EIDI	Emerging Infectious Disease Initiative
EIS	Epidemic Intelligence Service
ELISA	Enzyme-linked immunosorbent assay
EMCR	Electronic medical care record
EMS	Emergency Medical Service
EMT	Emergency medical technician
EPA	Environmental Protection Agency
ERDEC	Edgewood Research, Development and Engineering Center, U.S. Army

FABS	Force-amplified biological sensor
Fab	Antibody fragment
FBI	Federal Bureau of Investigation
Fc	Crystallizable fragment (of antibody)
FDA	Food and Drug Administration
FEMA	Federal Emergency Management Agency
FOWG	Fiber-optic evanescent wave guide
FTIR	Fourier Transform Infrared Spectrometry
GA	Tabun
GB	Sarin
GC/FTIR	Gas Chromatography Fourier Transform Infrared Spectrometry
GC/MS	Gas Chromatography Mass Spectrometry
GC-MS-MS	Gas Chromatography Tandem Mass Spectrometry
GD	Soman
GF	Cyclosarin
HAZMAT	Hazardous materials
HD	Sulfur mustard
HIV	Human immunodeficiency virus
HMT	Hexamethylene tetramine
HPAC	Hazard prediction and assessment capability
HPLC	High-performance liquid chromatography
HSEES	Hazardous substances emergency events surveillance
IDLH	Immediately dangerous to life and health
IMS	Ion mobility spectrometry
IND	Investigational new drug
IOM	Institute of Medicine
IPDS	Improved Chemical Agent Point Detection System
IU/L	International units per liter
JCAD	Joint Chemical Agent Detector
JCAHO	Joint Commission on Accreditation of Healthcare Organizations
JCBAWM	Joint Chemical Biological Agent Water Monitor
JLIST	Joint Service Lightweight Integrated Suit Technology
JPOBD	Joint Program Office for Biological Defense
JPOCD	Joint Program Office for Chemical Defense

JSLSCAD	Joint Service Lightweight Standoff Chemical Agent Detector
JUN	Junin virus
LANL	Los Alamos National Laboratory
LCR	Ligase chain reaction
LD_{50}	Dose lethal to 50 percent of the population exposed
LF	Lethal factor
LIDAR	Light detection and ranging
LLNL	Lawrence Livermore National Laboratory
MALDI-MS	Matrix-assisted laser desorption mass spectrometry
MANAA	Medical aerosolized nerve agent antidote
MARCORSYSCOM	Marine Corps Systems Command
MiniCAD	Miniature chemical agent detector
MMST	Metropolitan Medical Strike Team
NAME	Nitroarginine methylester
NARAC	National Atmospheric Release and Advisory Center
NBC	Nuclear, biological, chemical
NDA	New Drug Application
NDMS	National Disaster Medical System
NFkB	Nuclear factor-kappa B transcription factor
NFPA	National Fire Protection Association
NIAID	National Institute of Allergy and Infectious Diseases
NIH	National Institutes of Health
NIOSH	National Institute for Occupational Safety and Health
NIPAC	National Infrastructure Protection Center
NMRI	Naval Medical Research Institute
NOAA	National Oceanic and Atmospheric Administration
NRC	National Research Council or Nuclear Regulatory Commission
NRL	Naval Research Laboratory
NSWC	Naval Surface Warfare Center
OEP	Office of Emergency Preparedness
OP	Organophosphate
ORNL	Oak Ridge National Laboratory
OSHA	Occupational Safety and Health Administration
2-PAM	Pralidoxime chloride
PA	Protective antigen

PAHP	para-aminoheptanophenone
PAOP	para-aminooctanoylphenone
PAPP	para-aminopropiophenone
PAPR	Powered air purifying respirator
PBB	Polybrominated biphenyls
PCB	Polychlorinated biphenyls
PCC	Poison control center
PCR	Polymerase chain reaction
PDD	Presidential Decision Directive
PHS	Public Health Service
PID	Photo ionization detector
PIRS	Photoacoustic infrared spectroscopy
PPE	Personal protective equipment
ProMED	Program for Monitoring Emerging Diseases
PTSD	Post traumatic stress disorder
RBC	Red blood cell
RDEC	Research, development, and engineering center
RNA	Ribonucleic acid
RT	Reverse transcriptase
RVF	Rift Valley fever
SAW	Surface acoustic wave
SBIR	Small business innovative research
SciPUFF	Second-order Closure Integrated Puff
SCBA	Self-contained breathing apparatus
SCPS	Simplified Collective Protection System
SDA	Strand displacement amplification
SEB	Staphylococcal enterotoxin B
SFAI	Swept frequency acoustic interferometry
SOF	Special Operations Forces
SOPs	Standard operating procedures
STEPO	Self-contained Toxic Environment Protective Outfit
TAS	Transcription-based amplification system
TDG	Thiodiglycol
TOF-MS	Time-of-flight mass spectrometry
TSP	Topical Skin Protectant
TSWG	Technical Support Working Group
UAV	Unmanned aerial vehicle
UPT	Up-converting phosphor technology

USAMRICD	US Army Medical Research Institute of Chemical Defense
USAMRIID	US Army Medical Research Institute of Infectious Diseases
UV	Ultraviolet
VA	Veterans Affairs (Department of)
VEE	Venezuelan equine encephalomyelitis
VIG	Vaccinia-immune globulin
VX	Persistent nerve agent (o-ethyl S-[2-(diisopropylamino)ethyl]-methylphosphorofluoridate)
WEE	Western equine encephalomyelitis
WHO	World Health Organization
WMD	Weapons of mass destruction
WWW	World Wide Web
YF	Yellow fever

Contents

Chemical and Biological
TERRORISM

Executive Summary

The bombings of the World Trade Center in New York in 1993 and the Alfred P. Murrah Federal Building in Oklahoma City in 1995 have forced Americans to face the fact that terrorism is not something that happens only overseas. In addition, although the technology of producing and delivering chemical and biological weapons has existed for decades, the nerve gas attacks in Matsumoto in 1994 and Tokyo in 1995 by an apocalyptic religious cult and the subsequent revelation of the cult's attempts to acquire and use biological weapons have added a new dimension to plans for coping with terrorism. Traditional military approaches to battlefield detection of chemical and biological weapons and the protection and treatment of young healthy soldiers are not necessarily suitable or easily adapted for use by civilian health providers dealing with a heterogeneous population of casualties in a peacetime civilian setting.

For these reasons, the Institute of Medicine (IOM), in collaboration with the Commission on Life Sciences (CLS), was asked by the U.S. Department of Health and Human Services' Office of Emergency Preparedness (OEP) to:

- collect and assess existing research, development, and technology information on detecting potential chemical and biological agents and protecting and treating both the targets of attack and health care providers, and
- provide specific recommendations for priority research and development.

This report describes current civilian capabilities as well as ongoing and planned research and development (R&D) programs. It identifies some areas in which innovative R&D is clearly needed and assesses current R&D work for its applicability to coping with domestic terrorism.

ASSUMPTIONS

Pre-incident intelligence about specific agents will always be important, for it is not possible to be prepared for all possible agents in all possible circumstances. As a practical matter, the committee has taken as its reference point the relatively short list of chemical and biological agents that are discussed in the U.S. Army's handbooks for the medical management of chemical and biological casualties: nerve agents, cyanides, phosgene, and vesicants such as sulfur mustard; the bacteria-produced poisons staphylococcal enterotoxin B and the botulinal toxins; the plant-derived toxin ricin; the fungal metabolite T-2 mycotoxin; and the infectious microorganisms causing anthrax, brucellosis, plague, Q-fever, tularemia, smallpox, viral encephalitis, and hemorrhagic fever. As the body of the report notes, some are clearly more of a threat than others, and Appendixes C and D provide longer lists of potential chemical and biological agents, respectively, that have been assembled by other groups.

For the above agents, a particularly threatening means of delivery, on which both military offensive and protective programs and the committee's considerations have concentrated, is as vapors or aerosols designed to cause poisoning or disease as a result of inhalation. Nevertheless, it would be a mistake to assume that terrorists will not be able to use other agents, even novel ones, or other means of delivery, including contamination of food or water supplies.

As a practical measure, the committee chose to frame analyses of the possible utility of technology and R&D programs within three general scenarios. At one extreme is an overt attack rapidly producing significant casualties at a specific time and place—something similar to the Oklahoma City bombing, but involving a chemical or biological agent rather than, or in addition to, high explosives. Near the other extreme is a covert attack with an agent (for example, any of the bacteria or viruses alluded to in the previous paragraph) producing signs and symptoms in those exposed only after an incubation period of days or weeks, when the victims might be widely dispersed. The third scenario involves attempts at preemption, such as full-time monitoring of high-risk targets (e.g., the White House), deployment for specific events (Olympic Games, or the State-of-the-Union address), or simply dealing with a suspicious package.

The committee recognizes that for nearly any specific locale, a terrorist attack of any sort is a very low-probability event, and for that reason expensive or time-consuming actions in preparation for such events are extremely difficult for local governments to justify. Moreover, much of what could contribute to averting or mitigating casualties from terrorist chemical or biological attacks is urgently needed anyway to avert or mitigate severe hazards to health from toxic substances and prevailing or emerging infectious diseases of natural origin. As a result, the committee first gives special attention to developing recommendations for actions that will be valuable even if no attack ever occurs. A second type of recommendation focuses on specific actions that would be valuable in some of the more plausible scenarios. A third type of suggestion involves more generic, long-term research and development, although, even here, much of what needs to be done to deal with possible terrorist incidents will be of benefit to the nation's health irrespective of actual attack.

ORGANIZATION OF THE REPORT

This report analyzes preparedness at four levels of medical intervention—local emergency response personnel, initial treatment facilities, state departments of emergency services and public health, and a variety of federal agencies. The specific capabilities assessed are pre-incident intelligence (Chapter 2); detection and identification of chemical and biological agents in the environment and in clinical samples from victims (Chapters 4 and 6); personal protective equipment (Chapter 3); recognizing covert exposures of a population (Chapter 5); mass-casualty decontamination and triage procedures (Chapter 7); availability, safety, and efficacy of drugs, vaccines, and other therapeutics (Chapter 8); prevention and treatment of psychological effects (Chapter 9); and computer related tools for training and operations (Chapter 10). A list of specific R&D needs is provided at the end of each chapter. These R&D needs, numbering 61 in all, are summarized in eight overarching recommendations.

PRE-INCIDENT COMMUNICATION AND INTELLIGENCE

The response of even the most well prepared medical facilities will be markedly improved by advance notice from the law enforcement community. The latter understandably fear compromising ongoing investigations, but may not fully appreciate the substantial impact even very general information about possible incidents can have in facilitating a rapid and effective response by the medical community. Receipt of information concerning a possible mass-casualty event need not involve more than a few key individuals who can review the organization's seldom-used plan and begin to think about treatment options and where and how to obtain needed antidotes and drugs, make hospital beds available on short notice, ensure adequate staffing levels.

Recommendation 1. There needs to be a system in every state and major metropolitan area to ensure that medical facilities, including the state epidemiology office, receive information on actual, suspected, and potential terrorist activity.

Specific R&D needs:

• A formal communication network between the intelligence community and the medical community.
• A national mechanism for the distribution of clinical data to the intelligence and medical communities after an actual event or exercise.

PERSONAL PROTECTIVE EQUIPMENT

Personal protective equipment (PPE) refers to clothing and respiratory apparatus designed to shield an individual from chemical, biological, or physical hazards. The "universal precautions" (gloves, gown, mask, goggles, etc.) employed by medical personnel to prevent infections will generally provide protection from the biological agents under discussion, but it is difficult to say with confidence which, if any, civilian workers have suitable chemical PPE, because the testing and certification demanded by the Occupational Safety and Health Administration (OSHA) has not, until very recently, involved military nerve agents or vesicants, and military PPE that has been tested for protection against those agents generally does not have the testing and certification that would allow its use by civilian workers.

Hospitals receive not only field-decontaminated patients but also "walk-ins" who may have bypassed field decontamination. Despite Joint Commission on Accreditation of Healthcare Organizations standards calling for hospitals to have hazardous materials (Hazmat) plans and conduct Hazmat training, two recent reviews have suggested that most hospitals in the United States are ill prepared to treat chemically contaminated patients.

Recommendation 2. The committee endorses continued testing of civilian commercial products for suitability in incidents involving chemical warfare agents, but research is still needed addressing the bulk, weight, and heat stress imposed by current protective suits, developing a powered air respirator with greatly increased protection, and in providing detailed guidance for hospitals on dermal and respiratory protection.

Specific R&D needs:

- Increased protection factors for respirators.
- Protective suits with less bulk, less weight, and less heat stress.
- Evaluation of the impact of occupational regulations governing use of personal protective equipment.
- Uniform testing standards for protective suits for use in chemical agent incidents.
- Guidelines for the selection and use of personal protective equipment in hospitals.
- Alternatives to respirators for expedient use by the general public.

DETECTION AND MEASUREMENT OF CHEMICAL AND BIOLOGICAL AGENTS

Hazardous materials or "Hazmat" teams are routinely equipped with a variety of chemical detectors and monitoring kits, primarily employing chemical-specific tests indicating only the presence or absence of a suspected chemical or class of chemical. The most common detectors test for pesticides, chlorine, and cyanide, but not specifically for phosgene, vesicants, or nerve agents. Although chemical tests, detectors, and monitors used by the military are commercially available for civilian use, they have not been acquired by civilian organizations in appreciable numbers.

Laboratory assays indicating exposure to cyanide and anticholinesterase compounds such as nerve agents are known and available at many hospitals, but there is no current clinical test for mustard agents or other vesicants. However, for all of these agents except mustard, individuals receiving potentially lethal doses usually develop signs and symptoms within a matter of minutes after exposure. Therefore, initial diagnosis and treatment are likely to be based on observations of signs and symptoms by the paramedic or other health care professional on the scene.

Real-time detection and measurement of biological agents in the environment is more daunting, even for the military, because of the number of potential agents to be distinguished, the complex nature of the agents themselves, the myriad of similar microorganisms that are always present in the environment, and the impracticality of providing real time, continuous monitoring at even a fraction of the sites of potential concern. Few if any civilian organizations currently have, or can easily obtain, even a rudimentary capability in this area.

Some serological, immunological, and nucleic acid assays are available for identifying all of the biological agents being considered in this report, and many hospitals and commercial laboratories have the necessary equipment and expertise to perform these and similar assays. However, these diseases are extremely rare in the United States, and for that

reason these laboratories do not perform these assays regularly. It therefore seems unlikely that many labs will be immediately prepared to conduct the specific analytical test needed to confirm the presence of the
agent, even when the attending physician is astute enough to ask for the
appropriate test.

**Recommendation 3. The civilian medical community must find ways
to adapt the many new and emerging detection technologies to the
spectrum of chemical and biological warfare agents. Public safety and
rescue personnel, emergency medical personnel, and medical laboratories all need faster, simpler, cheaper, more accurate instrumentation for
detecting and identifying a wide spectrum of toxic substances, including but not limited to military agents, in both the environment and in
clinical samples from patients. The committee therefore recommends
adopting military products in the short run and supporting basic research necessary to adapt civilian commercial products wherever possible in the long run.**

Specific R&D needs:

• Evaluation of current Hazmat and EMS chemical detection equipment for ability to detect chemical warfare agents.
• Miniaturized and less expensive gas chromatography/mass spectrometry technology for monitoring the environment within fixed medical facilities and patient transport vehicles.
• Standard Operating Procedures for communicating chemical detection information from first responders to Hazmat teams, EMS teams,
and hospitals.
• Simple, rapid, and inexpensive methods of determining exposure
to chemical agents from clinical samples.
• Faster, cheaper, easier patient diagnostics that include rare potential bioterrorism agents.
• Inexpensive or multipurpose biodetectors for environmental testing and monitoring.
• Basic research on pathogenesis and microbial metabolism.
• Scenario-specific testing of assay and detector performance.

RECOGNIZING COVERT EXPOSURE IN A POPULATION

In the case of many biological agents, the time lag between exposure
to a pathogen and the onset of symptoms may vary from hours to weeks,
so effective response to a covert terrorist action will depend, not on fire
and rescue personnel, but upon (a) the ability of individual clinicians,

perhaps widely scattered around a large metropolitan area, to identify and accurately diagnose an uncommon disease or toxin response and (b) a surveillance system for collecting reports of such cases that is actively monitored to catch disease outbreaks as they arise.

The Centers for Disease Control and Prevention (CDC) operates a large number of infectious disease surveillance systems based on voluntary collaboration with state and local health departments, surveys, vital records, or registries. The best known of these systems, the National Notifiable Disease Surveillance System, currently includes several, but not all, of the diseases considered likely to be used in bioterrorism, and, like all passive surveillance systems, suffers from omissions and long-delayed reports. All of the systems depend upon confirmed diagnosis and are thus no help to a puzzled physician trying to arrive at a diagnosis. No federal funds are provided to state and local health departments to support these systems, and states' ability or willingness to support infectious disease surveillance has declined in recent years. CDC's Emerging Infections Program (EIP) is attempting to reverse this trend by making grants to state and local health departments for improving epidemiological and laboratory capability. Expanding the activities of these centers would be an excellent way to raise both the awareness of bioterrorism and the ability to respond to it.

In most plausible chemical terrorism scenarios, the rapid onset of toxic effects would lead to highly localized collections of victims within minutes or hours, so the need for active surveillance is less pressing. A network of regional poison control centers is well established, however, and, if its personnel were educated about military chemical weapons, would be well suited for surveillance. Poison control centers are also obvious candidates to serve as regional data and resource coordinating centers in incidents involving multiple sites or large numbers of patients.

Recommendation 4. Improvements in CDC, state, and local surveillance and epidemiology infrastructure must be undertaken immediately and supported on a long-term basis.

Specific R&D needs:

• Improvements in CDC, state, and local epidemiology and laboratory capability.
• Educational/training needs of state and local health departments regarding all aspects of a biological or chemical terrorist incident.
• Faster and more complete methods to facilitate access to experts and electronic disease reporting, from the health care provider level to global surveillance.

• Expanded pathogen "fingerprinting" of microbes likely to be used by terrorists and dissemination of the resulting library to cooperating regional laboratories.

• Symptom-based, automated decision aids that would assist clinicians in the early identification of unusual diseases related to biological and chemical terrorism.

MASS-CASUALTY DECONTAMINATION AND
TRIAGE PROCEDURES

The removal of solid or liquid chemical agent from exposed individuals is the first step in preventing severe injury or death. Civilian Hazmat teams generally have basic decontamination plans in place, though proficiency may vary widely. Very few teams are staffed, equipped, or trained for mass decontamination. Hospitals need to be prepared to decontaminate patients, despite plans that call for field decontamination of all patients before transport to hospitals. However, few hospitals have formal decontamination facilities; even fewer have dedicated outdoor facilities or an easy way of expanding their decontamination operations in an event involving mass casualties.

Recommendation 5. R&D in decontamination and triage should concentrate on operations research to identify methods and procedures for triage and rapid, effective, and inexpensive decontamination of large groups of people, equipment, and environments.

Specific R&D needs:

• The physical layout, equipment, and supply requirements for performing mass decon for ambulatory and nonambulatory patients of all ages and health in the field and in the hospital;

• A standardized patient assessment and triage process for evaluating contaminated patients of all ages;

• Optimal solution(s) for performing patient decon, including decon of mucous membranes and open wounds;

• The benefit vs. the risk of removing patient clothing;

• Effectiveness of removing agent from clothing by a showering process;

• Showering time necessary to remove chemical agents;

• Whether high-pressure/low-volume or low-pressure/high-volume spray is more appropriate for optimal cleaning of contaminated areas;

• The best methodology to employ in determining if a patient is "clean"; and

• The psychological impact of undergoing decontamination on all age groups.

AVAILABILITY, SAFETY, AND EFFICACY OF DRUGS, VACCINES, AND OTHER THERAPEUTICS

Vaccines against the agents of concern are, with only a couple of exceptions, of questionable utility, given the need to vaccinate far in advance of exposure. There are, in any case, licensed vaccines for only three of the biological agents being considered (anthrax, plague, and smallpox). There are few drugs of demonstrated effectiveness against any of the viral diseases of concern, nor are there safe and effective antitoxins to combat all the toxins on our short list (botulinum toxins A-F, SEB, ricin, and T-2 mycotoxin). Despite these shortfalls, given rapid response and/or accurate diagnosis, successful treatment of a very small number of individuals exposed to many of the chemical or biological agents is not beyond current medical capabilities. However, large numbers of casualties will quickly exhaust the limited supplies of antidotes, antibiotics, antitoxins, supportive medical equipment, and trained personnel that make that possible.

Recommendation 6. Conduct operations research on stockpiling and distribution of currently available antidotes for nerve agents and cyanide and give high priority to research on an effective treatment for vesicant injuries, investigation of new anticonvulsants and potential antibody therapy for nerve agents, development of improved vaccines against both anthrax and smallpox, development of a new antismallpox drug, and research on broad spectrum antiviral and novel antibacterial drugs.

Specific R&D needs:

• See Box 1 for a complete listing by agent and priority.

PREVENTION, ASSESSMENT, AND TREATMENT OF PSYCHOLOGICAL EFFECTS

Risks to victims and rescue and health care workers in such incidents include not only physical injury but psychological trauma. Research on post traumatic stress disorder (PTSD) has expanded far beyond studies of Vietnam veterans in the last 20 years, and includes a few studies of large-

BOX 1

R&D Needs in Availability, Safety, and Efficacy of Drugs and Other Therapies

HIGH PRIORITY

Nerve Agent
- Antidote stockpiling and distribution system
- Scavenger molecules for pretreatments and immediate post-exposure therapies

Vesicants
- An aggressive screening program focused on repairing or limiting injuries, especially airway injuries

Anthrax
- Vigorous national effort to develop, manufacture, and stockpile an improved vaccine

Smallpox
- Vigorous national effort to develop, manufacture, and stockpile an improved vaccine
- Major program to develop new antismallpox drugs for therapy and/or prophylaxis

Botulinum Toxins
- Recombinant vaccines, monoclonal antibodies, and antibody fragments

Non-specific Defenses Against Biological Agents
- New specific and broad-spectrum anti-bacterial and anti-viral compounds

MODERATE PRIORITY

Nerve Agents
- Intravenous or aerosol delivery of antidotes vs intramuscular injection
- Development of new, more effective anticonvulsants for autoinjector applications

Cyanide
- Dicobalt ethylene diamine tetraacetic acid, 4-dimethylaminophenol, and various aminophenones
- Antidote stockpiling and distribution system
- Risks and benefits of methemoglobin forming agents, hydroxocobalamin, and stroma free methemoglobin

Phosgene
- *N*-acetylcysteine and systemic antioxidant effects

Viral Encephalitides and Viral Hemorrhagic Fevers
- Antiviral drugs

Botulinum Toxins
- Botulinum immune globulin

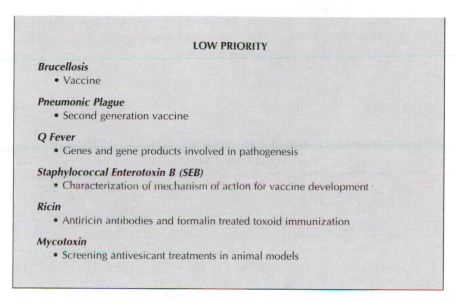

LOW PRIORITY

Brucellosis
- Vaccine

Pneumonic Plague
- Second generation vaccine

Q Fever
- Genes and gene products involved in pathogenesis

Staphylococcal Enterotoxin B (SEB)
- Characterization of mechanism of action for vaccine development

Ricin
- Antiricin antibodies and formalin treated toxoid immunization

Mycotoxin
- Screening antivesicant treatments in animal models

scale industrial accidents, among them, chemical spills. The latter studies have most often been epidemiological in nature, focusing on sequelae rather than treatment methods and their efficacy. A technique intended to prevent PTSD, Critical Incident Stress Debriefing (CISD), has gained wide acceptance among field emergency workers, and it can be expected that local police, fire, and emergency medical units will be familiar with the process and have plans to use it. Scientific evidence for its efficacy, however, is equivocal.

At the federal level, the National Disaster Medical Service (NDMS) includes special Disaster Medical Assistance Teams specializing in mental health, and the Federal Emergency Management Agency (FEMA) funds the Crisis Counseling Assistance and Training Program (CCP). Few practitioners have experience with chemical or biological disasters, however, and fewer still are knowledgeable about chemical or biological warfare agents.

Recommendation 7. Educational materials on chemical and biological agents are badly needed by both the general public and mental health professionals.

Specific R&D needs:

- Identify resource material on chemical/biological agents and enlist the help of mental health professional societies in developing a training program for mental health professionals

- Psychological screening methods for differentiating adjustment reactions after chem/bio attacks from more serious psychological illness.
- Evaluation of techniques for preventing or ameliorating adverse psychological effects in emergency workers, victims, and near-victims.
- Agent-specific information on risk assessment/threat perception by individuals and groups, and on risk communication by public officials.

COMPUTER-RELATED TOOLS FOR TRAINING AND OPERATIONS

This section of the report identifies relevant computer-related tools and pertinent health-effects information that could be used by medical and other first responders to train regularly or use operationally to enhance and sustain capabilities for identifying and managing chemical or biological terrorist incidents. These tools will also decrease the need for participation in large exercises that can be disruptive, logistically complicated, expensive, and sometimes unproductive.

Recommendation 8. The committee recommends support for computer software R&D in three areas: event reconstruction from medical data, dispersion prediction and hazard assessment, and decontamination and reoccupation decisions.

Specific R&D needs:

- Computer software for rapid reporting of unusual medical symptomology to public-health authorities and linking that data to both toxicological information and models of agent dispersion.
- Examination and field testing of current and proposed atmospheric-dispersion models to determine which would be most suitable for the emergency management community.
- Models of other possible vectors of dispersion (e.g., water, food, and transportation).
- Customizable simulation software to provide interactive training for all personnel involved in management of chemical or biological terrorism incidents.
- Information on the chemical, physical, and toxicological properties of the chemical and biological agents, in order to improve modeling of their environmental transport and fate and to better support recommendations on decontamination and reoccupation of affected property.

CONCLUSIONS

The recommendations listed above and the R&D needs associated with them are the true conclusions of the study. There are, nevertheless, some general conclusions that pervade the report as a whole, and it may be useful to make them explicit here. The most basic of these is that terrorist incidents involving biological agents, especially infectious agents, are likely to be very different from those involving chemical agents and thus demand very different preparation and response (the myriad of "chemical/biological" response teams being developed at federal, state, and local levels are, in fact, almost entirely focused on detection, decontamination, and expedient treatment of chemical casualties).

The second major conclusion that strongly influenced the committee's recommendations for research and development was the recognition that the military and civilian medical communities face very different situations with respect to prior knowledge about the identity of the enemy and the time and place of attack. Vaccination, for example, is an obvious preventive measure for a military force poised for combat against an enemy known or suspected to have a stockpile of certain biological weapons. The same holds true for deployment of chemical or biological detection systems and the use of highly specific antidotes and therapeutic and pre-treatment drugs: with reasonable intelligence about the enemy's capabilities and proclivities, these tools can be put into action rapidly and confidently. The value of all of these diminishes considerably in the most probable civilian terrorism situations, in which the enemy, the agent, the time, and the place of attack are unknown. This difference, even more than differences in the physiology and psychology of civilian and military targets, influenced the committee to emphasize treatment over prevention, broad-spectrum drugs, detection with familiar or multiagent equipment if possible, laboratory diagnostics based on commercial technology, decontamination without agent-specific equipment or solutions, modification of familiar or multipurpose protective clothing and equipment, and even the advisability of pre-hospital treatment. Chapter 2 argues for including the medical community in the distribution of pre-incident intelligence to maximize medical response in dealing with chemical or biological incidents, but, important as that is, the time scale envisioned in those arguments is much too short for truly preventive measures like vaccination or the introduction of unfamiliar specialized equipment.

Finally, for both chemical and biological incidents, there is an existing response framework within which modifications and enhancements can be incorporated. An attack with chemical agents is similar to the hazardous materials incidents that metropolitan public safety personnel contend with regularly. A major mission of public health departments is prompt identification and suppression of infectious disease outbreaks, and poison control centers deal with poisonings from both chemical and biological sources on a daily basis. The committee feels very strongly that it is important to make these existing mechanisms the focus of efforts to improve the response of the medical community to additional, albeit very dangerous, toxic or infectious materials.

1

Introduction

The bombings of the World Trade Center in New York in 1993 and the Alfred P. Murrah Federal Building in Oklahoma City in 1995 have forced Americans to face the fact that terrorism is not simply something that happens only overseas. Shocking as those attacks were to most Americans, the 1995 nerve gas attack on the Tokyo subway by an apocalyptic religious cult, Aum Shinrikyo, and the subsequent revelation of its attempts to acquire and use biological weapons (Broad, 1998) have added a new dimension to plans for coping with terrorism. The Tokyo attack, which killed 12 people and sent over 5,000 others to local hospitals, and an apparent rehearsal in the city of Matsumoto some months beforehand, were the first large-scale terrorist uses of a chemical or biological agent (Fainberg, 1997).

Scattered and smaller-scale incidents occurred previously (e.g., mercury poisoning of Israeli citrus in 1978, the Tylenol-cyanide poisoning of 1982 that led to current "tamper-proof" packaging, and the *Salmonella* poisonings by the Rajneesh cult in Oregon in 1984 intended to keep other voters from the polls), but a number of more recent incidents besides the Tokyo attack suggest that terrorists in the United States and abroad may be finding chemical and biological weapons increasingly attractive. In the United States, several members of a right-wing group called the Patriot's Council were convicted of acquiring the castor bean toxin ricin for use against local Minnesota officials in 1995. An Ohio man was arrested later that year and charged with fraudulently obtaining freeze-dried *Yersinia pestis* (plague) bacteria, and another individual was arrested in Arkansas in possession of a supply of ricin and castor beans and a collection of neoNazi books on making poisons. Overseas, German police confiscated a coded diskette containing directions for making mustard gas early in 1996, and political extremists in Tajikistan killed seven people and sickened a number of others with cyanide in 1995 (Oehler, 1996). The Aum Shinrikyo is reported to have experimented with anthrax and botulinum toxin before using the nerve gas sarin (GB) in the subway attack and may even have attempted to obtain a quantity of Ebola virus during the outbreak in Zaire (Fainberg, 1997; Broad, 1998).

The rapid breakup of the Soviet Union was accompanied by well publicized concern about the security of its nuclear arsenal. Other "weapons of mass destruction," namely, chemical and biological agents, drew less attention, but the extent of the Soviet chemical arsenal and the large Soviet biological weapons program are cause for concern about sales to or theft by terrorist groups and rogue states. Also disturbing is the fact that some chemical and biological agents and devices to deliver them efficiently can be inexpensively produced in simple laboratories or even legally purchased. Small quantities can cause massive numbers of casualties, covertly if the perpetrator so desires. The Tokyo attack, which may have been initiated prematurely because of justified suspicion that Japanese police were about to launch a preemptive strike, employed a very crude delivery system; otherwise, the number of deaths might have been far higher.

LEGISLATIVE BACKGROUND

The United States government, while continuing to pursue the goal of effective international prohibition of chemical and biological weapons through the Chemical Weapons Convention, the Biological Weapons Convention, and activities such as those of the Australian Group, has also recognized the need to address possible use of these agents by individuals or groups unlikely to be deterred by threats of economic sanctions or massive retaliation. In the past decade, Congress has passed three major laws aimed at preventing the acquisition and use of chemical or biological weapons by states, groups, or individuals. The Biological Weapons Act of 1989 makes it a federal crime knowingly to develop, manufacture, transfer, or possess any biological agent, toxin, or delivery system for use as a weapon. It calls for heavy criminal penalties on violators and allows the government to seize any such material for which no legitimate justification is apparent (P.L. 101-298). The Chemical and Biological Weapons Control Act of 1991 (CBWCA) established a system of economic and export controls designed to prevent export of goods or technologies used in the development of chemical and biological weapons to designated nations (P.L. 102-82).

The Anti-Terrorism and Effective Death Penalty Act of 1996 expanded the government's powers under CBWCA to cover individuals or groups who attempt or even threaten to develop or use a biological weapon. It also broadens the definition of biological agent to include new or modified agents produced by biotechnology and charges the Centers for Disease Control and Prevention (CDC) with creating and maintaining a list of biological agents that potentially pose a severe threat to public health and safety (P.L. 104-32). CDC is also charged with establishing regulations for the use and transfer of such agents that will prevent access to them by terrorists. CDC's new regulations, which took effect April 15, 1997, identify 24 microorganisms and 12 toxins, possession of which now requires registration with CDC and transfer of which now involves filing of forms by both shipper and receiver (Centers for Disease Control and Prevention, 1997). Appendix D is a list of those agents.

Another law, the Defense Against Weapons of Mass Destruction Act of 1996, directs the Secretary of Defense to take immediate actions to both enhance the capability of the federal government to respond to terrorist incidents and to support improvements in the capabilities of state and local emergency response agencies. In recognition of this requirement, an amendment (widely known as Nunn–Lugar II or Nunn–Lugar–Domenici after its congressional sponsors) to the Defense Authorization Act for Fiscal Year 1997 (P.L. 104-201) authorized $100 million to establish a military rapid response unit; to implement programs providing advice, training, and loan of equipment to state and local emergency response agencies; and to provide assistance to major cities in establishing "medical strike teams." The Department of Defense (DoD) has shared these funds with the Federal Emergency Management Agency (FEMA), the Federal Bureau of Investigation (FBI), the Department of Health and Human Services (DHHS), the Environmental Protection Agency (EPA), and the Department of Energy (DoE). Use of these funds for simple purchase of equipment for local users is, however, prohibited by the legislation.

Also relevant is the Local Firefighter and Emergency Services Training Act of 1996, which authorizes the Department of Justice, in consultation with FEMA, to provide specialized training to state and local fire and emergency services personnel.

In addition to congressional action, Presidential Decision Directive 39 (PDD-39), *United States Policy on Counterterrorism*, was issued in June, 1995. It specifies the responsibilities of federal agencies and their relationships to one another in the conduct of crisis management and consequence management. As defined in PDD-39, crisis management involves actions to anticipate and prevent acts of terrorism. United States law assigns primary authority for these actions, which are predominantly of a law enforcement nature, to the federal government, namely the FBI. Consequence management involves measures to protect public health and safety, restore essential government services, and provide emergency relief to governments, businesses, and individuals affected by acts of terrorism. United States law assigns primary authority in this sphere to the states; the federal government provides assistance as required. This assistance is coordinated by the FEMA, relying on procedures of the Federal Response Plan developed by 27 federal departments and agencies for responding to disasters of all kinds (Federal Emergency Management Agency, 1997).

PDD 62, *Combating Terrorism*, and PDD 63, *Critical Infrastructure Protection*, issued in May 1998, establish a National Coordinator for Security, Infrastructure Protection, and Counter-Terrorism and authorize the FBI to set up a National Infrastructure Protection Center (NIPAC) to issue warnings to public and private operators of essential elements of our government and economy. An additional element of the two directives is a four-part initiative focused on biological weapons. It calls for a national surveillance system based on the public health system, provision of local authorities with necessary equipment and training, stockpiles of vaccines and specialized medicines, and a research and development program on pathogen gene mapping to guide development of new and better medicines and vaccines.

CHARGE TO THE COMMITTEE

The Federal Response Plan designates the Secretary of DHHS, acting through the Assistant Secretary for Health, and the Office of Emergency Preparedness (OEP), to coordinate assistance in response to the public health, medical care, and health-related social service needs of victims of a major emergency and to provide resources when state and local resources are overwhelmed. DHHS's experience planning and preparing for possible terrorist actions aimed at the 1996 Atlanta Olympic Games and other events revealed that traditional military approaches to battlefield detection of chemical and biological weapons and the protection and treatment of young, healthy soldiers under relatively isolated and controlled circumstances are not necessarily suitable or easily adapted for use by civilian health providers dealing with a heterogeneous population of potential casualties in a civilian setting. The importance to terrorists of psychological impact, which may be significant even when the number of casualties is low, also suggests that a different approach from that of the military may be necessary. Advances in detection and personal protective equipment in the hazardous waste disposal and hazardous materials handling areas are also unlikely to be readily transferred to a mass-casualty situation requiring protection or extraction, decontamination, and treatment of large numbers of civilians of widely varying size, age, and health.

For these reasons, the Institute of Medicine (IOM), aided by the Commission on Life Sciences (CLS), was asked by OEP to conduct an 18-month study to (1) collect and assess existing research, development, and technology information on detecting chemical and biological agents as well as methods for protecting and treating both the targets of attack and the responding health care providers, and (2) provide specific recommendations for priority research and development. Areas of concern include, but are not limited to (1) the safety and efficacy of known and potential vaccines, prophylactic drugs, antidotes, and therapeutics; (2) vaccine production and distribution capabilities, surveillance for disease caused by biological agents, and real-time detection of chemical agents and rapid assays of biological agents; (3) the need for acute and chronic toxicological studies of emerging-threat agents; (4) plans for short-term and long-term follow-up of personnel exposed to chemical or biological agents; (5) adequacy and availability of personal protective equipment suitable for medical care providers; and (6) integrated triage, decontamination, and treatment practices and systems. The study thus focuses on medical responses to chemical or biological incidents and extends neither to prevention of terrorism nor to long-term actions like site remediation.

IOM and CLS assembled a committee of knowledgeable scientists and medical practitioners in accordance with established National Academy of Sciences procedures, including an examination of possible biases and conflicts of interest, and held an initial organizational and data-gathering meeting July 22–24, 1997. A roster with brief biographies of committee members is provided in Appendix A.

DATA COLLECTION

A wide variety of sources were used in assembling the requested inventory, which is attached as Appendix B. The initial meeting of the committee in July of 1997 provided an overview of important organizations and R&D programs within the federal government (Army, Navy, Marine and Defense Department units and laboratories, including the Defense Advanced Research Projects Agency; the Department of Energy's Chemical and Biological Nonproliferation Program; the Centers for Disease Control and Prevention; and the DHHS Office of Emergency Preparedness). Follow-up with the briefers provided a more detailed list of projects and points of contact for technical information. The Office of Emergency Preparedness shared information on promising technology from its files, and, of course, the committee members themselves contributed both personal contacts and specific information from their own files and experience. The World Wide Web provided much information about both relevant commercial products and R&D activity, and the following databases were accessed and searched: National Technical Information Service, Defense Technical Information Center, Federal Research in Progress, Federal Conference Papers, Medline, MedStar, and HSRProj.

Information on the products in the above inventory was distilled from a ProCite database of more than 450 records and entered into a series of databases, a description of which constitutes the gap and overlap analysis at the end of Appendix B. In the process, we eliminated most products or R&D that did not explicitly address military chemical or biological agents or appear to be sufficiently generic in nature to encompass those agents without a major change. Exceptions were made only in categories in which there were very few or no products or R&D explicitly directed at chemical and biological weapons. We also excluded technology represented in our database by only a single experiment, journal article, or Small Business Innovative Research contract (i.e., we focus on products and R&D *programs*).

ASSUMPTIONS AND PARAMETERS OF THIS REPORT

The committee's interim report (Institute of Medicine and National Research Council, 1998) focused on current civilian capabilities and made recommendations for action without evaluation of ongoing and planned research and development. That focus was selected primarily to provide a baseline against which to evaluate the utility of technology and R&D programs. The present report builds on that baseline by adding analyses of improvements in capability that might be possible through incorporation of technological innovations, either currently available or in some stage of research and development. However, the committee recognizes that technology is only one of a number of methods of improving civilian medical response and that it would be irresponsible to focus solely on technology, while ignoring potentially simpler, faster, or less expensive mechanisms, such as organization, manpower, training, and procurement. The committee's recommendations therefore include a number of suggestions for operational research in addition to calls for more traditional product development.

Chemical and Biological Agents Considered

Many of the actions required for effective consequence management are agent-specific (antidotes, for example). Some have argued that chemical and biological terrorism are especially vexing problems, because would-be terrorists have a much longer list of agents from which to choose than does a military force, which must be concerned with production in quantity, weaponization, storage, safety of their own personnel and civilian noncombatants, and contamination of desired physical and geographical objectives. Indeed, some have pointed out, correctly, that genetic engineering may eventually make the list of potential terror agents extremely long. In practice, the few chemical and biological terrorist incidents that have occurred to date have involved only a few different agents, and these agents are well known from military weapons programs. There is no guarantee that this will continue to be the case, indeed; it would be a grave mistake to assume that terrorists will not be able and willing to take advantage of biotechnology to produce new agents. Pre-incident intelligence about the specific agent suspected will always be important, for it is not possible to be prepared for all possible agents in all possible circumstances.

As a practical matter, the committee has taken as its reference point and as a framework within which to discuss current capabilities a limited number of agents: the relatively short list of chemical and biological agents discussed in the U.S. Army's recent textbook on medical aspects of chemical and biological warfare (Sidell et al., 1997). These are nerve agents, cyanide, phosgene, and vesicants, such as sulfur mustard; the bacteria-produced poisons botulinum toxin and staphylococcal enterotoxin B (SEB); the plant-derived toxin ricin; the fungal metabolite T-2 mycotoxin; and the infectious microorganisms causing anthrax, brucellosis, plague, Q-fever, tularemia, smallpox, viral encephalitis, and hemorrhagic fever. More comprehensive lists, including these chemical agents and the CDC list of restricted biological agents, are provided as Appendixes C and D respectively, to illustrate the breadth of the problem facing planners.

Terrorism Scenarios Considered

The committee also recognizes that terrorist incidents can take a wide variety of forms. Evaluation of civilian medical and public health capabilities and shortfalls will be scenario dependent. Like the number of possible agents, the number of possible scenarios is also very large. Important variables include the extent of prior intelligence or warning about the time, place, or nature of the attack; the degree to which time and place of the attack itself is obvious; and the number and location of individuals exposed.

Again, as a practical measure, the committee chose to limit analyses of the possible utility of technology and R&D programs to three general scenarios. The first is an overt attack producing significant casualties at specific time and place—something similar to the Oklahoma City bombing, but involving chemical or biological agent rather than, or in addition to, high explosives. A second, quite different scenario is a covert attack with an agent producing disease in those exposed only after an incubation period of days or weeks, when the victims might be widely dispersed. The infectious agents listed in the previous section have incubation periods ranging from 2 days to 6 or 8 weeks, depending on the agent and the dose received (Franz et al., 1997). A third scenario involves attempts at preemption, such as full-time monitoring of likely targets (the White House, subways), deployment for specific events (Olympic Games or the State-of-the-Union address), or simply dealing with a suspicious package. Consequence management in these three general scenarios is obviously quite different, qualitatively and quantitatively: there will be no 911 call to which emergency personnel respond, indeed no site or identifiable event in the second scenario, so training and equipment for that scenario must be focused not on public safety personnel (fire and rescue, emergency medical services) but on hospitals, medical laboratories, and public health officials.

For the above agents and scenarios, a particularly threatening means of delivery, on which both military offensive and protective programs and the committee's considerations have concentrated, is as vapors or aerosols designed to cause poisoning or infectious disease as a result of inhalation. Nevertheless, it would be a mistake to assume that terrorists will not be able to use other agents, even novel ones, or other means of delivery, including product tampering, attacks on crops, and contamination of food or water supplies. In fact, the IOM, in conjunction with the National Research Council Board on Agriculture, recently released a report assessing the current food safety system and providing recommendations on scientific and organizational changes needed to ensure an effective system for present and future generations (Committee to Ensure Safe Food from Production to Consumption, 1998). Although focused on naturally occurring contamination rather than deliberate acts, carrying out the report's recommendations is likely to be an important first step in guarding the country against chemical or biological attack by this route as well.

Constraints on Local Resources

The charge to the committee focuses on what is possible and desirable. It makes no reference to cost, financial or otherwise. The committee nevertheless recognizes that for nearly any specific locale (with the possible exception of a few obvious areas like Washington, D.C.) a terrorist attack of any sort is a very low-probability event, regardless of the magnitude of the consequences, and that taking expensive or time-consuming actions in preparation for such events is extremely difficult for state and local governments to justify. Moreover, many of the actions that could contribute to averting or mitigating casualties from terrorist chemical or biological attacks are urgently needed anyway to avert or mitigate severe hazards to health from toxic substances and prevailing or emerging infectious diseases of natural origin. As a result, the committee has given special attention to actions that will be valuable even if no terrorist attack ever occurs. A second type of recommendation focuses on very specific actions that would be valuable in a few more plausible terrorist scenarios. A third category of suggestions involves more generic, long-term research and development programs. Even here, much of what needs to be done to deal with possible terrorist incidents will be of benefit to the nation's health irrespective of actual attack.

CURRENT CIVILIAN CAPABILITIES

The committee's Interim Report (Committee on R&D Needs for Improving Civilian Medical Response to Chemical and Biological Terrorism Incidents, 1998) focused on current capabilities. Much of that analysis survives in the present document, albeit modified in a number of places due to new developments or simply receipt of additional information. Table 1-1 is a summary table adapted from a similar table in the interim report. The leftmost column of the table lists capabilities or actions likely to be required for an effective response to the medical consequences of a chemical or biological incident. The table entries represent the committee's best estimates of current capabilities, though there are few hard data to support them and the committee recognizes that capabilities vary widely at the state and local level. In effect, the table also provides an outline of the remainder of the report, which will describe current preparedness in each of these areas, along with possible improvements achievable through existing technology or research and development. Regardless of technology, of course, integrated planning and coordination among different levels of the medical community will be necessary for effective response.

For purposes of this report, we differentiate four levels of medical intervention, primarily on the basis of proximity to the precipitating event or initial victims. Response to a distinct, immediately recognizable terrorist incident (as opposed to a covert release of an agent whose effects would not be apparent for hours or days) would, in most instances, be initiated by law enforcement or fire and rescue personnel, followed at some point by a hazardous materials (Hazmat) team and emergency medical technicians. This is the group referred to in Table 1-1 as "Local Responders." In the same table, "Initial Treatment Facilities" refers to the fixed-site medical facilities to which victims might initially be transported (or transport themselves) or that might initially be called upon for assistance by victims or personnel on the scene. Under "State" in the table, we

TABLE 1-1 Relative Capabilities for Response to Civilian Chemical and Biological Incidents at Four Levels of Medical Care

Capability	Local Responders		Initial Treatment Facilities		State		Federal	
	Chemical	Biological	Chemical	Biological	Chemical	Biological	Chemical	Biological
Receipt of pre-incident intelligence	L	L	L	L	S	S	S	S
Detection, identification, and quantification of agents in the environment	S	L	L	L	S	S	H	S
Personal protective equipment	S	S	L	S	L	S	S	S
Safe and effective patient extraction	S	S	N/A	N/A	N/A	N/A	S	S
Methods for recognizing symptoms and signs in patients	S	S	S	S	L	L	S	S
Detection and measurement of agent exposure in clinical samples	L	L	L	S	L	S	H	H
Methods for recognizing covert exposure in populations	N/A	N/A	S	S	S	S	S	S
Mass-casualty triage techniques and procedures	S	S	S	S	L	L	S	S
Methods/procedures for decontamination of exposed individuals	S	S	L	L	L	L	S	S
Availability, safety, and efficacy of drugs and other therapies	L	L	S	S	L	L	S	S
Prevention, assessment, and treatment of psychological effects	S	S	S	S	S	S	S	S

NOTE: H = highly capable; S = some capability; L = little or no capability; and N/A = not applicable.

refer primarily to state departments of emergency services and public health and to regional resources, such as poison control centers and public health laboratories. A state public health agency would probably initiate the systematic response to a covert release of an agent with delayed effects (e.g., anthrax).

The "Federal" category in Table 1-1 refers to capabilities that are many and varied. Upon request from the governor, the Federal Emergency Management Agency (FEMA) may deploy an emergency response team, the health and medical services portion of which is the responsibility of the Department of Health and Human Services (DHHS), specifically the Office of Emergency Preparedness. The DHHS National Counterterrorism Plan includes initiatives both to create or improve local capabilities and to enhance the existing National Disaster Medical System (NDMS). One initiative involves organizing, equipping, and training groups of local fire and rescue personnel as Metropolitan Medical Strike Teams (MMST) in 25 or more of the nation's largest cities. The first of these teams, in Washington, D.C., became operational in early 1997, and contracts have been awarded to establish 10 more teams in 1998. The goal of these teams is to enhance local planning and response systems capability, tailored to each city, to care for victims of a terrorist incident involving a weapon of mass destruction (nuclear, chemical, or biological, although, in practice, the core of most of the teams is the Fire Department and its hazardous materials team, with a resulting emphasis on chemical attack). This is to be accomplished by providing special training to a subset of local emergency responders (120 to 300, depending on the size of the metropolitan area); specialized protective, detection, decontamination, communication, and medical equipment; special pharmaceuticals and other supplies; and enhanced emergency medical transport and emergency room capabilities. Other capabilities include threat assessment, public affairs, epidemiological investigation, expedient hazard reduction, mental health support, victim identification, and mortuary services. Twenty-five of these teams will obviously cover only a small proportion of the U.S. population, but they should also serve as effective test beds and models for Hazmat teams of the future.

The National Disaster Medical System (NDMS) supplements state and local medical resources by delivering direct medical care to disaster victims. Disaster Medical Assistance Teams (DMAT) provide prehospital treatment. Sixty existing teams, some tailored to focus on pediatrics, burns, mental health, and other specialties, including mortuary services, are in place around the country. Like military reserve units, the teams are community-based and composed of local health providers who train on weekends. Twenty-one are fully deployable and can be on the scene in 12 to 24 hours with enough food, water, shelter, and medical supplies to remain self sufficient for 72 hours and treat about 250 patients per day. Three teams are being organized and trained specifically to respond to chemical or biological terrorism. NDMS hospitalization assistance is accomplished though a regional network of 72 Federal Coordinating Centers that are run by the Department of Veterans Affairs (VA) and the DoD. These centers have agreements with private sector hospitals to make ready a total of more than 100,000 in-patient hospital beds; the VA provides medicines and DoD provides patient transportation.

The Centers for Disease Control and Prevention (CDC) is a widely recognized source of expertise in the diagnosis of infectious agents known to be pathogenic in humans, and their epidemiological and laboratory resources are often called upon to assist state health departments identify and manage outbreaks of severe unexplained illness. In cases of suspected biological or chemical terrorism CDC itself can consult with experts at academic institutions and research institutes and several DoD medical research units specializing in biological or chemical defense: the U.S. Army Medical Research Institute of Infectious Diseases (USAMRIID), the U.S. Army Medical Research Institute of Chemical Defense (USAMRICD), the U.S. Navy Medical Research Institute (NMRI), and the U.S. Navy Environmental and Preventive Medicine Unit.

Representatives of these units and additional DoD organizations with expertise in bomb disposal (the Army's 52nd Ordnance Group) and the detection and disposal of chemical and biological weapons (the Army's Technical Escort Unit and Chemical Treaty Laboratory and the Naval Research Laboratory) form the DoD Chemical/Biological Rapid Response Team (C/B-RRT). The C/B-RRT is a deployable source of advice and expertise that can coordinate more extensive and more specialized assistance as necessary. Under some circumstances, a U.S. Marine Corps unit called the Chemical Biological Incident Response Force (CBIRF) may provide assistance in evacuation, decontamination, and medical stabilization of victims. This 350-person force is based at Camp LeJeune, North Carolina and can have an advance party airborne four hours after notification. Other deployable units designated to assist local civilian responders include Specialty Response Teams at the Army's Regional Medical Centers, which can provide advice on casualty management and coordinate more extensive support. An Aeromedical Isolation Team of physicians, nurses, and technicians from USAMRIID specializes in the transport of patients with highly contagious diseases. One destination for those patients may be a small isolation ward at USAMRIID designed for the care of patients requiring the highest levels of containment.

A newly established hotline to the U.S. Army Chemical Biological Defense Command provides 24-hour access to these DoD assets. Given the very rapid action of chemical weapons, telephonic advice will probably be the most valuable assistance that local authorities can count on. CBIRF and other "hands-on" units are likely to play a major role only when deployed in advance to a site where there is reason to suspect an attack (e.g., the 1996 Atlanta Olympics). The Secretary of Defense has announced plans to expand and decentralize assistance to local, state, and federal agencies responding to attacks with chemical or biological weapons by equipping and training elements of the National Guard units (under the control of state governors) and the Army and Air Force Reserves to provide decontamination, medical care, security, and transportation. Initial actions would establish rapid assessment and initial detection (RAID) teams in 10 areas designated by FEMA, to be followed by 55 reconnaissance elements and 127 decontamination teams throughout all 50 states. Expertise and equipment associated with these teams, like the CBIRF, will be heavily biased toward response to chemical attack, an event at which, also like CBIRF, barring predeployment, they will almost certainly arrive too late to provide significant help with the medical response.

Capabilities at each level and promising leads for improvements are treated in detail in the following chapters, but several general conclusions that emerge serve to summarize the reasoning behind the ratings of Table 1-1.

First, the committee believes that the incubation period associated with infectious biological agents makes responding to an attack with such agents very different from responding to a chemical attack. Thus, victims of biological attack may not be concentrated in time and space the way victims of most chemical attacks will be. Recent emphasis on preparing "local responders" for chemical and biological terrorism is therefore primarily preparation for chemical attacks.

Second, in many of the areas surveyed in subsequent chapters, we note that some capability, often quite good, exists for incidents involving a small number of victims. Regardless of preparation, there will be some unpreventable casualties in all but the most incompetent attacks, but without planning, education, supplies, equipment, and training, the casualty count will mount rapidly when the number of persons exposed escalates, particularly as the event is likely to be unprecedented in a community. For both chemical and biological exposures however, there is an existing response framework within which modifications and enhancements specific to chemical and biological terrorism can be incorporated. An attack with chemical agents is similar to the hazardous materials incidents that metropolitan public safety personnel contend with regularly. A major mission of public health departments is prompt identification and suppression of infectious disease outbreaks, and poison control centers deal with poisonings from both chemical and biological sources on a daily basis. It would be a serious tactical and strategic mistake to ignore (and possibly undermine) these long-neglected mechanisms in efforts to improve the response of the medical community to additional, albeit very dangerous, toxic materials. It would be similarly ill advised to ignore the existing mechanisms for providing federal disaster assistance to local communities.

Local governments and hospitals are reluctant to spend large amounts of money and time preparing for what they judge as unlikely events. Federal organizations can, therefore, be very important. This is particularly true in the case of biological agent incidents, where onset of signs or symptoms is delayed, variable, and potentially continuing, and victims may be widely dispersed. The National Disaster Medical System (NDMS), for example, would be a critical component of response to any large-scale biological attack. The NDMS might also serve a useful role in a large-scale chemical attack, though the rapid onset of effects from these agents puts a premium on actions within the first few hours following exposure. For that reason, properly trained and organized Metropolitan Medical Strike Teams organized by local communities with Public Health Service funding may be the most useful federal help in managing the medical consequences of a chemical attack. Similar help from deployable military teams will be optimal only if intelligence allows for predeployment or the attack occurs near the team's home base.

Finally, it will be apparent that federal regulations intended to protect the public in very different circumstances may have, in fact, impeded efforts to prepare for chemical or biological terrorism. Regulations on worker safety apparel, for example, have made it difficult for civilian rescue workers to take advantage of military equipment specifically designed for protection from chemical warfare agents. Similarly, the difficulty of obtaining the human efficacy data currently required for FDA approval of specific treatments for chemical and biological warfare agents may limit their use in mass-casualty situations. Furthermore, in the case of many treatments, collection of the data on efficacy necessary for full FDA approval will not be possible for ethical reasons or economically attractive to a potential manufacturer because of limited market potential.

2

Pre-Incident Communication and Intelligence: Linking the Intelligence and Medical Communities

The threat of chemical and biological terrorism, coupled with current world events, has caused the many disciplines responsible for the health and welfare of the public to evaluate their ability to respond adequately to an intentional use of a weapon of mass destruction. The national medical community—including public health agencies, emergency medical services, hospitals, and health care providers—would bear the brunt of the results of a chemical or biological attack. An attack of a chemical or biological agent could result in civilian mortality and morbidity that have not been seen in natural disasters or infectious outbreaks in the United States since the influenza epidemic of 1918–1919.

As noted in the preceding chapter, the medical community must prepare for three general types of incidents. The first is an overt release resulting in a chemical or biological exposure to a population with its subsequent morbidity or mortality. In most cases illness exposure or risk is known from the moment the exposure is identified, and efforts to mitigate its effects as well as treat victims can begin immediately. The second type of terrorist incident is a covert release involving an agent with a delayed onset of illness and delayed identification that a population is at risk. In this situation, exposure, illness, or injury may be widespread before mitigation or treatment can begin. Finally, there is the threat of a release by a terrorist group that has identified itself or is discovered through normal intelligence operations. In this case, medical authorities can serve as a surveillance system for law enforcement by looking for medical indicators suggestive of terrorist activity. Such indicators might include unusual illness or injury in a community. Although medical organizations have historically not been recipients of pre-incident intelligence, this practice needs to change in light of recent concerns about chemical and biological terrorism.

The CDC now maintains a database of individuals and organizations possessing any of 36 biological agents (listed in Appendix D) with potential to cause a severe threat to public safety and health. The legislation does not require CDC to share this information with state or local health departments, however, and sharing has not been done in any systematic way. Although facilities willing to report to CDC that they are working with these agents are unlikely to be terrorist threats themselves, they may be targets of terrorists, victims of theft by rogue employees, or the source of an unintended release. All of these events will be handled better if the local medical community is aware of the possibility.

Of far more importance is the need for an institutionalized linkage between the law enforcement and medical communities. The response of even the most well prepared medical facilities will be markedly improved by advance notice from the law enforcement community. The latter understandably fear compromising ongoing investigations, but may not fully appreciate the substantial impact even very general information about possible incidents can have in facilitating a rapid and effective response by the medical community. Receipt of information concerning a possible mass-casualty event need not involve more than a few key individuals who can review the organization's seldom-used plan and begin to think about treatment options, where and how to obtain needed antidotes and drugs, make hospital beds and resources available on short notice, and ensure adequate staffing levels. Inclusion of these key medical personnel in anti-terrorist intelligence activity would no doubt be facilitated by their willingness to undergo training on the needs of the law enforcement community, especially procedures for proper preservation of evidence.

After-action reports on the Tokyo subway incident (Obu, 1996; Olson, 1996; Yanagisawa, 1996) provide an example of the value of communication between the law enforcement and medical communities as well as an example of a missed opportunity for communication within the medical community that might have made the medical response even more effective than it was. Japanese police had apparently been planning a raid on Aum Shinrikyo facilities throughout Japan, and for that reason the government had ordered medical supplies, including nerve agent antidotes, not normally stocked in quantity by hospitals (anonymous comments in Obu, 1996). One of the reasons for the raids was the suspected involvement of the Aum Shinrikyo in a previous toxic gas incident in the city of Matsumoto almost a year before the Tokyo attack (Morita et al., 1995). The release in that city in 1994 of what was subsequently identified as sarin resulted in seven deaths and the treatment of an additional 250 people. A group of Matsumoto physicians, recognizing that data from humans exposed to sarin were very rare, collected a great deal of information on these patients, which they sent to Tokyo hospitals and the Ministry of Health and Welfare as soon as they heard of the subway attack. Although the information reportedly was helpful, it seems obvious that a more formal mechanism by which the Matsumoto group could have more rapidly and systematically alerted other cities and hospitals to such an unusual event might have been even more valuable.

In this country, the District of Columbia's Emergency Management Office and Public Health Agency were provided an extensive, although generic, briefing on the terrorism threat just before the start of the Gulf War. Similar briefings have no doubt taken place on occasions such as the 1996 Atlanta Olympic Games, and personal relationships may provide good communication between the law enforcement and medical communities in some cities. However, few have the sort of structural links that the MMSTs are attempting to build into their operations—a law enforcement section, headed by a local law enforcement officer, one of whose major duties is to establish relationships with the local FBI office and other law enforcement agencies sufficient to ensure that the team has the maximum prior warning of potential nuclear, chemical, or biological incidents.

It is necessary to have an accurate ongoing assessment and prioritization of the chemical and biological agents that pose the greatest threat as well as identification of the agents that pose the most credible threat (using some of the 36 agents on the CDC list of restricted agents). In order for the medical community to efficiently prepare and respond to chemical and biological terrorism, it must be equipped with the latest and most accurate information on current risks. This is essential to ensure adequate preparatory measures, such as stocking and maintaining appropriate and sufficient amounts of vaccines, antibiotics, and other pharmaceutical agents and to ensure maximum effort in providing for the safety of health care providers, paraprofessionals, and support personnel. These events often involve the use of medications or vaccines that are often not available in large enough supply locally and, even if maintained in regional stockpiles, still require time to obtain or produce adequate stores to effect meaningful treatment or prophylaxis.

Emergency medical workers would benefit from preincident warnings by entertaining a broader range of hypotheses when entering an illness or injury site where the risks are unclear. These added considerations could be the difference between the loss or incapacitation of the rescue team or the secondary spread of potentially contaminated/infected patients to other health care sites.

Identification of the agents used in chemical and biological terrorism may involve sophisticated tests that take several hours to days for results. Initial signs and symptoms in victims of biological terrorist events may present as common disease processes. Not thinking about the possibility of these more lethal or infectious diseases runs the risk of secondary spread and additional cases of epidemic proportions. Early warning of potential threats will stimulate an earlier screen of potentially exposed individuals for intoxication or infection and a more rapid public health response. This is even more important if multiple infectious agents or a combination of chemical and infectious agents are suspected. Traditional medical teaching is to try to explain a clinical condition by a single disease process. Containing the disease outbreak or the chemical contamination is the most important public health responsibility of consequence management of a biological or chemical terrorist event. Timely and accurate pre-incident intelligence is essential to achieve this goal.

Health officials are often the first medical personnel to be contacted by the press whenever an epidemic or other public health threat occurs. Early knowledge of the threat of a chemical or biological event would allow public health officials to develop plans for effective risk communication and ensure appropriate coordination with law enforcement authorities. Accurate and timely information from public health officials is essential to prevent public panic. Benefits of effective communication include: reducing the inappropriate use of scarce health care resources by low-risk individuals and ensuring that individuals at highest risk present for treatment.

Although further research is needed into the best ways to improve communication, the many advantages of providing the medical community information obtained by agencies monitoring and gathering intelligence on terrorist activity and threats is vital.

In summary, the medical community has the diversity to respond to a wide array of biological and chemical health emergencies, including those which are intentional. Although the intelligence community has a legitimate need to protect its sources and the law enforcement community its operations, current and accurate information must be made available to the medical community in the pre-incident phase as well as the response phase of an event. This includes any information regarding credible threats to a community and potential agents that might be used. Information on successful interdictions, including agents and plans for their dissemination, would even be valuable after the fact for planning and training. The bilateral sharing of information, intelligence, and clinical data will ensure that victims receive the most efficient care possible, based on fact and experience rather than on assumptions and conjecture.

R&D NEEDS

The committee is encouraged by the recent provisions of Presidential Decision Directive 63 that call for establishing a national center to warn owners and operators of critical economic and governmental infrastructures of terrorist threats. We hope that as the details of this center are developed, the medical community will not be overlooked. To enhance communication to and within the national medical community, the following R&D needs have been identified:

2-1 Development of a formal communication network between the intelligence community and the medical community that incorporates local emergency management agencies as an important element and thereby creates a mechanism for public health and emergency management officials to gain access to intelligence information. This might best be accomplished by incorporating public health and other health professionals into the intelligence community to monitor and assess biological agents and terrorist threats from the perspective of a health emergency (health intelligence liaisons). Operations research should be done to identify what triggers should initiate the transfer of intelligence information and what medical and intelligence agencies should be involved.

2-2 Development of a national mechanism for the distribution of clinical data, including treatment modalities, patient outcomes, efficacy of pharmaceutical agents, best practices, and other medical interventions, to the intelligence and medical communities after an actual event or exercise.

3

Personal Protective Equipment

The term PPE (Personal Protective Equipment) refers to clothing and respiratory apparatus designed to shield an individual from chemical, biological, and physical hazards. This chapter includes a description of the types of PPE that address the needs of emergency workers, health care providers, and potential victims of terrorist attacks. It notes the general lack of access of many health care providers and potential victims to any kind of PPE. It also addresses the lack of specific regulatory standards for commercial PPE for use against military agents. Finally, the chapter discusses recently completed, ongoing, and planned research and development programs focused on PPE appropriate for response to terrorist attacks.

The chapter focuses primarily on protection from chemical agents, in part because of the fact that protection from hazardous chemicals will generally provide protection against biological agents as well, and in part because of the committee's belief that, by and large, biological agent incidents are not likely to be evident until well after release of the agent, at which point most agent not already in victims will have dissipated or degraded.

TYPES OF PPE AND REGULATORY STANDARDS

The amount and type of protection required in any hazardous materials incident depends upon the hazard and the duration of exposure anticipated, but a National Institute for Occupational Safety and Health

(NIOSH)/OSHA/EPA classification system is often used in describing general levels of protection:

- *Level A* provides maximal protection against vapors and liquids. It includes a fully encapsulating, chemical-resistant suit, gloves and boots, and a pressure-demand, self-contained breathing apparatus (SCBA) or a pressure-demand supplied air respirator (air hose) and escape SCBA.
- *Level B* is used when full respiratory protection is required but danger to the skin from vapor is less. It differs from Level A in that it incorporates a nonencapsulating, splash-protective, chemical-resistant suit (splash suit) that provides Level A protection against liquids, but is not airtight.
- *Level C* utilizes a splash suit along with a full-faced positive or negative pressure respirator (a filter-type gas mask) rather than an SCBA or air line.
- *Level D* is limited to coveralls or other work clothes, boots, and gloves.

OSHA requires Level A protection for workers in environments known to be immediately dangerous to life and health (i.e., where escape will be impaired or irreversible harm will occur within 30 minutes), and specifies Level B as the minimum protection for workers in danger of exposure to unknown chemical hazards. The NIOSH and the Mine Safety and Health Administration designate performance characteristics for respirators and provide approval for all commercially available respirators. Chemical protective clothing is not subject to performance standards established by a government agency, but the American Society for Testing and Materials (ASTM) has developed methods for testing the permeability of protective clothing materials against a battery of liquids and gases. The National Fire Protection Association (NFPA) has incorporated the ASTM test battery into the currently accepted standards for protective suits for hazardous chemical emergencies. Although a basic rule in selecting PPE is that the equipment be matched to the hazard, none of the ASTM permeability tests employ military nerve agents or vesicants. However, the NFPA is currently in the process of developing testing standards that will address the threat of nerve agents, cyanides, and vesicants.

ACCESS TO PPE

In the event of a chemical-agent incident, it is most likely that the first emergency personnel on the scene will be police or firefighters. The former will almost never have chemical PPE and should simply relay observations to the latter. Firefighter "turnout" or "bunker" gear designed for fire and heat resistance provides only minimal protection against hazardous chemicals, but firefighters often have sufficient respiratory protection (SCBA) available to allow for a rapid extraction and initial decontamination of victims at a location away from the primary source.

Hazmat teams have a small number of Level A suits. Recent data from tests of 12 different suits from six manufacturers by the Army's Domestic Preparedness Program (Belmonte, 1998) indicate that many commercially available Level A suits may provide good protection against nerve and mustard agents. The duration of protection will vary with individual fit, activity level, concentration of agent, and exposure route. Most, if not all, other NFPA-certified commercial Level A suits are likely to provide protection from most concentrations for at least brief periods. The current focus of Hazmat team activity is, in fact, short-term operations to control or mitigate a release, rather than sustained efforts locating and extracting victims or doing site remediation.

Emergency medical personnel most often have Level C PPE if they have any at all. This is likely to be appropriate for treatment of decontaminated victims, but unless the agent can be identified and its concentration established as nonlife-threatening, OSHA regulations would call for Level B protection.

A similar situation exists at local hospitals that may receive not only field-decontaminated patients but also "walk-ins," who may have bypassed field decontamination. Some authors have argued that Level C protection or even Level D protection (hospital gown, goggles, surgical mask, and latex gloves) is adequate for emergency department personnel; others argue for a universal PPE policy that will cover the exceptional cases (e.g., Level B in all cases until thorough decontamination is completed). Although the Joint Commission on Accreditation of Healthcare Organizations has established standards for hospitals calling for hazardous material plans and training, it does not specify details of either, and two recent reviews have suggested that most hospitals in the United States are ill prepared to treat contaminated patients (Cox, 1994; Levitin and Siegelson, 1996). A 1989 study of 45 California hospitals found that only two of the 45 actually had any personnel protective equipment assigned to the emergency department, and one of those two kept it in an ambulance that was not always at the hospital (Gough and Markus, 1989).

POTENTIAL ADVANCES

In the event of a chemical or biological terrorist act, there is a need to protect two main populations—the responders/health care providers, and the victims. The following section will review potential advances and R&D needs for both of these groups.

Responders and Health Care Providers

Military PPE has been tested for protection against chemical weapons agents (e.g., one or more nerve agents, mustard, and lewisite), but generally does not have the certification by NIOSH or NFPA that would allow its purchase and use for any purpose by civilian workers. Some progress in addressing this impasse was made in conjunction with the Army's Chemical Stockpile Emergency Preparedness Program (CSEPP). In order to make recommendations for civilian emergency responders in communities adjacent to chemical weapons stockpiles (Argonne National Laboratory, 1994; Centers for Disease Control and Prevention, 1995a), CSEPP sponsored tests of commercial respirator filters (Battelle Laboratories, Inc., 1993), fabrics used in commercial chemical suits (Daugherty et al., 1992), and one commercially available splash suit (Arca et al., 1996) using nerve and mustard agents. Subsequent U.S. Army testing of four Level A suits, four Level B suits, and four Level C suits has resulted in approval of two Level A commercial suits for use in chemical agent emergencies at Army facilities and purchase of commercial Level A and Level C PPE by MMSTs (United States Army Chemical Demilitarization and Redemption Activity, 1994).

As a result of testing undertaken by the CSEPP program, a number of filter canisters for powered air purifying respirators (PAPRs) were shown to provide protection against exposure to chemical weapons agents. PAPRs allow greater mobility than SCBAs and might support responders performing decontamination and medical triage and treatment. However, in order to meet regulatory requirements, responders can use PAPRs only in an environment in which the level of exposure to chemical weapons agents can be measured. Not only must monitors be available, they must detect the chemicals at appropriate concentrations. The necessary concentrations are determined by the effectiveness of the respirator (designated by the protection factor assigned to the class of respirator) and the acceptable exposure limit for the contaminant. Protection factors are a measure of performance based on a ratio of the contaminant concentration outside the mask to the concentration inside the mask. The airborne exposure limit (AEL) is an 8-hour time weighted average of exposure, for a 40 hour work week. PAPRs as a class (at the air flow rate tested), are assigned a protection factor of 50 by NIOSH. PAPRs therefore can be used only when monitors can detect the chemical at a concentration fifty times the AEL for that chemical. Lack of practical monitoring equipment (easily used in the field) that can detect chemical weapons agents at the limits required results in difficulties meeting current regulatory requirements. Therefore, although PAPRs might provide adequate protection against exposure to chemical weapons agents for some responders, SCBAs or in-line respirators are required to meet regulatory standards.

New insights into respirator design in the last few years have resulted in the development of improved respiratory protection. Protection factors appear to increase by an order of magnitude with a switch from a facemask to a hood design. Combining the hood-style mask with a blower unit has achieved even more significant results. One such mask currently in development under the U.S./Israel Agreement on Cooperative Research and Development Concerning Counter-Terrorism takes advantage of these combined technologies. The hood-style blower system achieved protection factors of 50,000 in the preliminary test results reviewed and is being designed for chemical/biological protection. The hood style also has the advantage of being a one-size-fits-all system. Continued efforts to develop this respirator technology and obtain regulatory approval for civilian emergency responders should be supported.

Cutaneous exposure to toxic liquids during a terrorist incident is a concern (the hazard to skin from vapor exposure is likely to be low). Protective suits tested against chemical weapons agent simulant (see above) are similar in basic design to those routinely used by civilian responders. The suits have some modifications, such as specially sealed seams, and are more expensive than similar suits that are not approved for use against chemical weapons agents. The suit testing program for commercial suits used the criteria identified for the military Joint Service Lightweight Integrated Suit Technology (JLIST) program, and results indicated that the commercial suit tested provided protection greater than the military's Battle Dress Overgarment (BDO).

Problems with suits remain and can include bulk, weight, and durability. Heat stress is a significant problem. The military has fielded suits with advantages over BDOs (for example SARATOGA system clothing). JLIST technology currently being fielded is expected to provide chemical/flame protection, reduced heat stress, increased durability, and the ability to be washed. Suits under development in the Advance Lightweight Chemical Protection program may offer even greater improvements. The program is based on selectively permeable membrane technology, and suits are expected to be lighter in weight, less bulky, and result in less thermal stress that JLIST garments. The National Aeronautics and Space Program has developed a prototype Level A suit as part of their Global NBC Emergency Response Technology Program that uses cryogenic air to provide for suit cooling as well as a larger air supply. Preliminary testing has suggested significant improvements in both heat management as well as work period efficiency and duration in simulated hazardous materials incidents.

Protection for Possible Victims of Terrorist Attack

Pocket-sized masks intended for victim rescue and self-rescue during chemical and biological incidents are available from several manufacturers (for example, Fume Free, Essex, Giat). One system uses layers of activated charcoal cloth to remove chemical toxicants and a particulate element for particle removal. Testing has shown the system to be effective against nerve agent simulant, hydrogen cyanide, and tear gas. The one-size fits all mask uses a hood design with a neck seal.

Equipment intended for use by the public was available and was used in Israel during the Gulf War. Improper respirator use resulted in some deaths (Hiss and Arensburg, 1994). Additional data were gathered on issues such as the psychological response of civilians during respirator use and physiological effects in children (Arad et al., 1994). Respirators intended for the public were subsequently redesigned to prevent accidental death and improve the efficacy and comfort of the equipment. Equipment developed in Israel included: hood-style mask and blower systems for civilian emergency responders; one size fits all hood and blower systems for adults and for children ages 3–7; and portable infant protection cribs.

Collective shelter is an alternative to individual protective equipment, but would be most useful in situations in which a specific group has been targeted (e.g., use by the military). Systems are available from a few manufacturers. Sheltering in place is an option for the general population when evacuation is not practical. UNOCAL Corporation, a large manufacturer of agricultural chemicals, has developed simple kits containing tape, plastic, a shelter-in-place video, and symptom cards and has distributed the kits to families and schools near their processing plants. Families can use the kits to quickly seal a room as a shelter. The approach appears cost effective. Other programs have been developed to educate the public about sheltering, which, in some circumstances, if undertaken too late or carried out improperly, may actually increase the exposure of those in the shelter. The "Wally Wise Guy" program, developed in Texas and now used in several other states, educates children about sheltering in place. Communities in Oregon and Washington near a chemical weapons stockpile have jointly developed an educational video on sheltering in place.

Areas in Need of Further Work

Generally, a range of equipment is available to protect both emergency responders and the general public during chemical and biological events. However, problems remain. One is lack of uniform testing standards for suits. Others, such as the potential for heat stress in many ensembles, should be a priority in current development programs. A possibility being pursued by the U.S. military, for example, is the use of selectively permeable membrane technology.

One of the most significant problems remaining is the choice of respiratory protection by responders. This choice is inextricably linked to availability of monitors capable of measuring toxicants at levels that satisfy regulatory requirements. Without adequate monitoring equipment responders are limited to working in Level A PPE. This limitation imposes unacceptable training burdens and expense on many agencies. Level A ensembles also result in limited stay times where the toxic agent or its concentration is unknown, difficulties in treating patient due to the bulk of the ensemble, and greater potential for heat stress. It would be an advantage for health care providers to use Level B and C PPE. As discussed above, some Level C respirators (PAPRs) have been tested and would provide protection against chemical/biological agents. However, current chemical field detectors have inadequate sensitivity to support use of PAPRs in chemical agent incidents. Two approaches could mitigate the situation within the current regulatory framework. Fielding respirators with greatly increased protection factors, hood and blower systems for example, would raise the concentration level at which use of an SCBA is mandated. Current gross level monitors may then be adequate. The second approach is to increase the sensitivity of field detectors provided to responders so that the appropriate level of PPE can be chosen with confidence. A third approach would be to reassess current regulations for the occupational use of PPE, regulations that do not apply to the general public, in the specific context of emergency response situations. For example, current regulatory standards that are protective for chronic occupational exposures might be reviewed and special criteria developed. When the criteria are met hospital staff remote from an incident could potentially use PAPRs (supported by gross level monitoring) for short-term exposures.

Respiratory protection used in Israel for general public has been improved as the result of insights gained during the Gulf War. However, the risk to the public from chemical exposure needs to be balanced against the risks of respirator use from erroneous use. Improper use can result not only in loss of protection against the chemical but also in injury from use of the respirator itself, particularly in individuals with asthma or other respiratory disorders. Adequate warning time would be needed to distribute respirators to the public and to carefully educate people on their use. Hood and blower systems offer less risk to the public than other respirator systems and are available for adults and children. Respirator systems based on a facemask rather than a hood system would need to be fit for each individual to ensure a tight seal. These respirators can not be issued to citizens with facial hair. Blower units supply filtered air to the hood, thereby eliminating the need for the individual to actively pull air through the filter and also ensuring that minor or temporary breaks in the hood-neck seal result in filtered air leaving the hood rather than contaminated air being drawn in. Batteries used in blower systems would require monitoring and eventual replacement, however, which increases the difficulty in maintaining a program for the general population. It is unlikely that early intelligence would be adequate to support distribution of respirators and education of the recipients. This and the other considerations discussed here make the option of providing respirators to the general public less attractive than either evacuation or some form of sheltering in place.

PPE Specifically for Biological Agents

Terrorist use of a biological agent presents very different needs for and uses of personal protective equipment than use of a chemical agent. Unless pre-incident intelligence leads responders to an incident prior to release of a biological agent, the majority of terrorist scenarios would likely involve a covert release of agent. Since most of the biologic agents have incubation times ranging from hours to weeks between exposure and manifestation of clinical symptoms, the majority of the biological agent aerosol is likely to have dissipated from the area of release prior to recognition by first responders that a biological incident has occurred.

With the exception of smallpox virus and to a lesser extent plague bacteria, person-to-person transmission of these diseases rarely occurs if "universal precautions" are maintained (e.g., gloves, gown, mask, and eye protection). The majority of infected patients can be cared for without specialized isolation rooms or specialized ventilation systems. Cohort nursing with the usual practice of universal precautions will provide adequate protection. The hemorrhagic virus infections may be transmissible via a respirable aerosol of blood—respiratory protection of workers caring for these patients is required.

In the event that pre-incident intelligence puts fire and rescue personnel at the scene of a release, the same PPE they would employ for a chemical incident should serve to protect them from biological agents as well. Most of the infectious agents and toxins are most efficiently delivered as a respirable aerosol, so respiratory protection would be the primary means of protection from these agents. This can be accomplished by either self-contained supplied air breathing devices (SCBA) or high-efficiency particle respirators (HEPA filters). Eye protection and protective clothing sufficient to provide a barrier will protect from cutaneous infection with these agents. An exception to these biological protective equipment strategies is T-2 mycotoxin, which requires an approach similar to chemical agents (Wannamacher et al., 1991; Wannamacher and Weiner, 1997).

Protection of the first responders will likely involve barrier protection similar to the equipment currently used for potentially infectious patients supplemented by SCBA or HEPA filters. It is important to note the current OSHA regulations for response workers require protection levels similar to those required for chemical agents. These regulations should be reevaluated for applicability in light of the risks posed by biological toxins.

Implementing these PPE strategies may prove difficult, as it is human nature to proceed to maximum protection when the perceived danger is unknown or unusual. It is important to emphasize basic principles of infectious disease control and emphasize the lack of person-to-person transmission for the majority of the biological agents when responding to such incidents, so as to maximize the available medical resources to provide care for the largest number of victims.

R&D NEEDS

Research and development in personal protective equipment has yielded vastly improved protection for the military and, to some extent, civilian first responders. However, the use of even the most up-to-date respirator is greatly restricted by the necessity of air monitoring, time of exposure limitations, and relatively low protection factors. Civilian first responders are also hampered by the weight, size, and heat of the protective suits. Aside from issues surrounding the equipment itself, policy and regulation also influence use and effectiveness of personal protective equipment. As listed below, the committee recommends that research and development continue to focus on better and more effective equipment, but also recommends that current policy and procedures be reviewed as well.

3-1 Continue research on new technologies that increase protection factors achieved by respirators.

3-2 Continue research on chemical protective suits that better address issues of bulk, weight, and heat stress.

3-3 Evaluate current occupational regulations governing use of personal protective equipment in the context of terrorism.

3-4 Support current efforts to develop uniform testing standards for protective suits intended for use in response to terrorism.

3-5 Research and develop recommendations for selection and use of appropriate personal protective equipment in hospitals.

3-6 Research alternatives to respirators for expedient use by the general public (e.g., sheltering in place, ventilation system filters) in terrorist incidents.

4

Detection and Measurement of Chemical Agents

Rapid identification of the chemical or biological agents involved in any hazardous material (Hazmat) incident is vital to the protection of first responders and emergency medical personnel at local medical facilities as well as to the effective treatment of casualties. This chapter of the report deals with devices for detecting and identifying chemical agents and is followed by two chapters focusing on biological agents. However, a potential complication that can easily be overlooked is the possibility that a terrorist attack may involve the use of more than a single agent. Therefore, detection of one agent should not bring identification efforts to a premature halt. Instead, detection of any agent should be taken as an indication of an imminent threat and should therefore provoke more extensive testing.

CHEMICAL WARFARE AGENTS IN THE ENVIRONMENT

When addressing the requirement for chemical agent detection at the scene of a terrorist incident, it is important to consider who really has the responsibility for detection operations. Is it the police, the fire department, the Hazmat team, or the EMS units? If the responsibility falls on medical assets, how will the use of detection equipment increase the efficiency of the health care team in preserving life and preventing further injury? Furthermore, will the absence of more sensitive equipment somehow inhibit their performance? These are important questions that must be addressed by policymakers and incident response planners.

Because of their physical properties, CWA use in a domestic terrorist incident may not be associated with a high-explosive event. Rather, these agents may well be dispersed in a manner that would involve a vapor hazard within a confined space. The type of incident seen in Tokyo, while minimized by the inefficient release of the CWA, is an excellent example of the type of incident expected as a result of terrorist use of CWA. The highest probability of detecting the presence of CWA occurs in cases in which there is a continuing source of vapor. By the time emergency medical personnel arrive at an incident, inevitably the agent will have dispersed significantly. In the case of cyanide and phosgene, and most nerve agents, detection in the environment may not be possible by the time monitoring equipment is in place at the scene. In fact, once casualties of a vapor CWA incident are outside the area of the attack and accessible to medical personnel, the signs and symptoms of the patients may be the only detection method available, and the threat of spread of the CWA hazard from casualties may be minimal. However, in the case of the Tokyo sarin attack, it has been reported that up to 9 percent of EMS workers and a significant number of hospital staff experienced acute symptoms of nerve agent toxicity due to exposure to casualties in unventilated areas (Okumura et al., 1998b). Given the increased threat of CWA terrorism and the various CW agents that can be used, emergency responders must have accurate and timely detection information or the ability to detect and identify a CWA at the time of their response. Again in the case of the Tokyo subway attack, the first identification made was inaccurate, and it was not until three hours post incident that accurate detection of sarin was made and the information disseminated. The medical personnel on site will require equipment capable of detecting the widest range of chemical agents. For medical personnel, detection equipment may include rapid, minimally invasive or noninvasive clinical assays for various chemical agents or for the effects of the chemical agents, that is, cholinesterase inhibition. Without this ability, more individuals may be exposed, including emergency response and hospital personnel attempting to care for casualties. Chemical agent detection will be an essential part of both medical crisis and consequence management. Detection and identification of the chemical agent or agents at the scene of a terrorist incident must not be accomplished at the expense of rapid and appropriate medical treatment of chemical casualties.

An emergency response incident that involves the release of any chemicals or toxic materials will typically be categorized as a hazardous materials (Hazmat) incident. The response to a Hazmat incident is somewhat standardized across the country. Specialized Hazmat teams are normally called in to address these situations. Hazmat teams are typically part of the fire services and will possess a majority of the locality's chemical detection equipment. The emergency responders who arrive on the scene first, however, must be capable of determining that a Hazmat incident has occurred. These first responders will be the individuals responsible for determining whether the Hazmat team should be called for assistance. Most emergency response vehicles do not carry any chemical detection equipment.

Chemical detection equipment currently used by Hazmat teams varies considerably by locality. For large metropolitan areas, current detectors range from adequate instrumentation to absolutely no capability for CWA detection. Hazmat teams are routinely equipped with a variety of chemical detectors and monitoring kits, primarily chemical-specific tests indicating only the presence or absence of the suspected chemical or class of chemical. A negative response of the test means only that a specific substance is not present in significant quantity; a positive response says nothing about the possible presence of other hazardous agents. Colorimetric tubes, designed for the detection of known and unknown gases, are commonly used by Hazmat teams. There are over 200 different tubes available that can detect a variety of chemicals. The tubes used by responders are sold in basic detection sets typically consisting of the tubes and a hand or mechanical pump. The pump, used to draw the air sample through the tube, will simultaneously carry out a volume measurement with each stroke. Direct reading detector tubes can be used for both short-term measurement and long-term measurement. The long-term tubes consist of two types: one requiring a constant flow pump, and the other a diffusion detector tube. Hazmat team analytical capabilities commonly include tests for chlorine, cyanide, phosgene gas, and organophosphate pesticides. The last of these tests may respond to the military nerve agents, but the requisite validation studies have not been conducted. Colorimetric tubes for the detection of CWA are not standard issue items for Hazmat teams and rarely include a means of detecting the chemical vesicants, such as sulfur mustard or Lewisite.

Many modern detection devices used by Hazmat teams have not been thoroughly tested for their utility and reliability to detect CWA. There is an ongoing effort under the sponsorship of the U.S. Army-managed Domestic Preparedness Program to test currently used Hazmat detection systems against classical CW agents. Preliminary analysis, for example, has shown that combustible gas indicators and pH paper, which are available to most Hazmat teams, will not serve as CWA detectors. The combustible gas detector was designed to detect and measure concentrations of combustible gases and/or vapors in the air, such as carbon monoxide, oxygen, and hydrogen sulfide, while pH paper is simply too generic to be useful as an indicator of CWA.

Currently Available Detection Technologies

A wide variety of detection equipment is available commercially and through the Department of Defense (DoD). Tests, detectors, and monitors of varying sensitivity (lowest level detectable) and specificity (ability to distinguish target from similar compounds) have been developed and/or used by the armed forces to identify the nerve agents and vesicants (Table 4-1). A comparison of the column labeled "Sensitivity" with the data of Table 4-2 reveals that for all but the most expensive of these devices, the sensitivity of most currently available Army systems is adequate for detection of the presence of immediately dangerous concentrations of chemical agents, but too low for them to be appropriate to ensure the complete health and safety of victims and responders. Many currently fielded Army chemical agent detection systems also suffer from excessive false positive alarms, a characteristic which is highly undesirable in a domestic civilian situation, especially in monitoring applications (as opposed to testing for the cause of signs and symptoms of poisoning in on or more patients). Current chemical detection technologies have been incorporated into mobile or stationary detection platforms and can be used as a point-source detector or as a remote (stand-off) detector. The primary differences between mobile detectors and stationary detectors are size, weight, portability, and logistical support requirements. The following section briefly describes the technological basis of this equipment, and Appendix B provides an extensive but perhaps not all-inclusive list of manufacturers that employ these technologies in their instruments.

Ion Mobility Spectrometry (IS) operates by drawing air at atmospheric pressure into a reaction region where the constituents of the sample are ionized. The ionization is generally a collisional charge exchange or ion-molecule reaction, resulting in formation of low-energy, stable, charged molecules (ions). The agent ions travel through a charged tube where they collide with a detector plate and a charge (current) is registered. A plot of the current generated over time provides a characteristic ion mobility spectrum with a series of peaks. The intensity (height) of the peaks in the spectrum, which corresponds to the amount of charge, gives an indication of the relative concentration of the agent present. This IMS technology is mainly used in mobile detectors to detect nerve, blister, and blood agents.

Electrochemical sensors function by quantifying the interaction between an analyte's molecular chemistry and the properties of an electrical circuit. Fundamentally, electrochemistry is based on a chemical reaction that occurs when the CWA enters the detection region and produces some

TABLE 4-1 Military Detection and Monitoring Equipment

Equipment	Agent	Sensitivity	Time	Cost	Operations/Maintenance/Limits	Notes
M-8 Paper	Nerve-G Nerve-VX Mustard-H Liquids only	100-μ drops 100-μ drops 100-μ drops	≤30 sec	$1 per book of 25 sheets	Disposable/hand-held Dry, undamaged paper has indefinite shelf life	Chemical agent detector paper; 25 sheets/book and 50 booklets/box; potential for false positives.
M-9 Paper	Nerve-G Nerve-VX Mustard-H Liquids only	100-μ drops 100-μ drops 100-μ drops	≤20 sec	$5 per 10-m roll	Disposable/hand-held 3-year shelf life Carcinogen	Adhesive-backed dispenser roll or books.
M-18A2 Detector Kit	Nerve-GB Nerve-VX Mustard-H, HN, HD, HT Lewisite-L, ED, MD Phosgene-CG Blood-AC Liquid, vapor, aerosol	0.1 mg/m^3 0.1 mg/m^3 0.5 mg/m^3 10.0 mg/m^3 12.0 mg/m^3 8.0 mg/m^3	2–3 min	$360	Disposable tubes Hand-held	25 tests per kit; Detector tubes, detector tickets, and M-8.

Continued on next page

TABLE 4-1 *Continued*

Equipment	Agent	Sensitivity	Time	Cost	Operations/ Maintenance/Limits	Notes
M-256A1 Detector Kit	Nerve-G and VX Mustard-HD Lewisite-L Phosgene oxime-CX Blood-AC, CK Vapor or liquid	0.005 mg/m^3 0.02 mg/m^3 2.0 mg/m^3 9.0 mg/m^3 3.0 mg/m^3 8.0 mg/m^3	15 min Series is longer AC— 25 min	$140	Disposable Hand-held 5-year shelf life	Each kit contains 12 disposable plastic sampler-detectors and M-8 paper.
M-272 Water Test Kit	Nerve-G and VX Mustard-HD Lewisite Hydrogen cyanide	0.02 mg/l 2.0 mg/l 2.0 mg/l 20.0 mg/l	7 min 7 min 7 min 6 min	$189	Portable/lightweight 5-year shelf life USN, USMC	Used to test raw or treated water; Type I and II detector tubes, eel enzyme detector tickets; Kit conducts 25 tests for each agent.
CAM Chemical Agent Monitor	Nerve-GA, GB, VX Blister-HD and HN Vapor only	0.03 mg/m^3 0.1 mg/m^3	30 sec ≤1 min	$7,500	Hand-held/portable battery operated 6–8 hours continuous use. Maintenance required.	Radioactive source. False alarms to perfume, exhaust paint, additives to diesel fuel.
ICAM Improved Chemical Agent Detector	Nerve-G and V Mustard-HD	0.03 mg/m^3 0.1 mg/m^3	10 sec 10 sec	$7,500	4.5 pounds Minimal training	Alarm only; False positives common.

	Agents	Concentration	Response time	Cost	Physical	Comments
ICAM-APD Improved Chemical Agent Detector—Advanced Point Detector	Nerve-G Nerve-V Mustard-H Lewisite-L	0.1 mg/m³ 0.04 mg/m³ 2.0 mg/m³ 2.0 mg/m³	30 sec 30 sec 10 sec 10 sec	$15,000	12 pounds including batteries Low maintenance Minimal training	Audible and visual alarm.
ICAD Miniature Chemical Agent Detector	Nerve-G Mustard-HD Lewisite-C Cyanide-AC, CK Phosgene-CG	0.2–0.5 mg/m³ 10 mg/m³ 10 mg/m³ 50 mg/m³ 25 mg/m³	2 min (30 sec for high levels) 2 min 15 sec	$2,800	8 oz pocket-mounted 4 months service No maintenance Minimal training	Audible and visual alarm; Marines; No radioactivity.
M-90 D1A Chemical Agent Detector	Nerve-G, V Mustard Lewisite Blood Vapor only	0.02 mg/m³ 0.8 mg/m³ N/A	10 sec 10 sec 80 sec	$16,000	15 lb. with battery Radioactive source exempt from licensing. Minimal training	Ion mobility spectroscopy and metal conductivity technology can monitor up to 30 chemicals in parallel. Alarm only.
M-8A1 Alarm Automatic Chemical Agent Alarm	Nerve-GA, GB, GD Nerve-VX Mustard-HD Vapor only	0.2 mg/m³ 0.4 mg/m³ 10 mg/m³	≤2 min ≤2 min ≤2 min	$2,555	Vehicle battery operated Maintenance required	Radioactive source (license required); Automatic unattended operation; Remote placement.
MM-1 Mobile Mass Spectrometry Gas Chromatograph	20-30 CWA Vapor	<10 mg/m² of surface area	≤45 sec	$300,000 military $100,000 civilian	Heater volatizes surface contaminants.	German "Fuchs" (FOX Recon System/Vehicle)

Continued on next page

TABLE 4-1 *Continued*

Equipment	Agent	Sensitivity	Time	Cost	Operations/ Maintenance/Limits	Notes
RSCAAL M-21	Nerve-G Mustard-H Lewisite-L Vapor	90 mg/m^3 2,300 mg/m^3 500 mg/m^3		$110,000	Line-of-sight dependent 10 year shelf life 2-person portable tripod	Passive infrared energy detector 3 miles; Visual/audible warning from 400 meters
SAW Mini-CAD	Nerve-GB Nerve-GD Mustard-HD Vapor	1.0 mg/m^3 0.12 mg/m^3 0.6 mg/m^3	1 min 1 min 1 min	$5,500	Minimal training Field use 1 pound No calibration	Alarm only; False alarms from gasoline vapor, glass cleaner.
ACADA (XM22)	Nerve-G Mustard-HD Lewisite Vapor	0.1 mg/m^3 2 mg/m^3 —	30 sec 30 sec —	$8,000	Vehicle mounted, battery powered Radioactive source (license required) Minimal training	Audible alarm; Bargraph display—low, high, very high.
Field Mini-CAMs	Nerve-G, V Mustard-H Lewisite-L	<0.0001 mg/m^3 <0.003 mg/m^3 <0.003 mg/m^3	<5 min <5 min <5 min	$34,000	Designed for field industry monitoring (10 lb.) 8 hours training 24 hour/7 day operations	Plug-in modules increase versatility; Threshold lower than AEL.

Instrument	Agents	Detection Limit	Time	Cost	Requirements	Comments
Viking Spectratrak GC/MS	Nerve-G, V Mustard-HD Many others	<0.0001 mg/m^3 <0.003 mg/m^3	<10 min <10 min	$100,000	Field use, but 85 pounds Needs 120v AC, helium 40 hours training	Lab quality analysis; Library of 62,000 chemical signatures.
HP 6890 GC with flame photometric detector	Nerve-G, V Mustard-HD Many others	<0.0001 mg/m^3 <0.0006 mg/m^3	<10 min <10 min	$50,000	Not designed for field use Gas, air, 220v AC 40 hours training	State-of-the-art gas chromatograph; Used by CWC treaty lab.

TABLE 4-2 Estimated Human Exposure Guidance for Selected
Chemical Warfare Agents

Agent	Vapor (mg/m^3)		Liquid ED$_{50}$ (mg/70 kg)
	AEL	EC$_{50}$	
Tabun (GA)	0.0001	< 1.7	< 880
Sarin (GB)	0.0001	< 0.8	1,000
Soman (GD)	0.00003	< 0.8	200
VX	0.00001	< 0.3	< 2.5
Sulfur mustard (HD)	0.003	3.33	600

AEL: The maximum airborne exposure concentration for an 8-hr workday.
EC$_{50}$: The airborne concentration sufficient to induce severe effects in 50% of those exposed for 30 minutes.
ED$_{50}$: The amount of liquid agent on the skin sufficient to produce severe effects in 50% of the exposed population.

SOURCES: (AEL) Edgewood Safety Office, 1996; (ED$_{50}$) National Research Council; (EC$_{50}$) derived from Ect$_{50}$ of Committee on Toxicology, 1998.

change in the electrical potential. This change is normally monitored through an electrode. A threshold concentration of agent is required, which corresponds to a change in the monitored electrical potential. This sensor technology provides a wide variety of possible configurations. Electrochemical detectors are used in mobile detectors to detect blister, nerve, blood, and choking agents.

In *flame photometry*, an air sample is burned in a hydrogen-rich flame. The compounds present emit light of specific wavelengths in the flame. An optical filter is used to let a specific wavelength of light pass through it. A photosensitive detector produces a representative response signal. Since most elements will emit a unique and characteristic wavelength of light when burned in this flame, this device allows for the detection of specific elements. Flame photometric detectors are commonly used in gas chromatographs.

Thermoelectric Conductivity. The electrical conductivity of certain materials can be strongly modulated following surface adsorption of various chemicals. Heated metal oxide semiconductors and room-temperature conductive polymers are two such materials that have been used commercially. The change in sensor conductivity can be measured using a simple electronic circuit, and the quantification of this resistance change forms the basis of sensor technology. Thermoelectric conductivity detection technology has only recently been applied to chemical agent detection.

Infrared Spectroscopy. The infrared (IR) region of the electromagnetic spectrum between 2.5 and 25 micrometers has proven valuable for the identification and quantification of gaseous molecular species. When infrared radiation passes through a gas, adsorption of radiation occurs at specific wavelengths that are characteristic of the vibrational structure of the gas molecules. Infrared detectors are used in mobile detectors to detect blister and nerve agent vapors.

Photoacoustic IR Spectroscopy. As in infrared spectroscopy, PIRS uses selective adsorption of infrared radiation by the CWA vapors to identify and quantify the agent present. A specific wavelength of infrared light is pulsated into a sample through an optical filter. The light transmitted by the optical filter is selectively adsorbed by the gas being monitored, which increases the temperature of the gas as well as the pressure of the gas. Because the light entering the cell is pulsating, the pressure in the cell will also fluctuate, creating an acoustic wave in the cell that is directly proportional to the concentration of the gas in the cell. Two microphones mounted inside the cell monitor the acoustic signal produced and send results to the control station. PIRS technology is fairly new, and it is expected that most agents can be detected with this technology.

Photo Ionization Detectors (PIDs) operate by passing the air sample between two charged metal electrodes in a vacuum that are irradiated with ultraviolet radiation, thus producing ions and electrons. The negatively charged electrode collects the positive ions, thus generating a current that is measured using an electrometer-type electronic circuit. The measured current can then be related to the concentration of the molecular species present. PID's are used in mobile detectors to detect nerve, blister, and mustard agents.

Surface Acoustic Wave (SAW) sensors detect changes in the properties of acoustic waves as they travel at ultrasonic frequencies in piezoelectric materials. The basic transduction mechanism involves interaction of these waves with surface-attached matter. Multiple sensor arrays with multiple coatings and pattern recognition algorithms provide the means to identify agent classes and reject interferant responses that could cause false alarms. Acoustic wave sensors are used in mobile detectors to detect nerve and blister agents.

Color-Change Chemistry. This technology is based upon chemical reactions that occur when CW agents interact with various solutions and substrates. The most common indicator (for a positive response) is a color change. Detection tubes, papers, or tickets use some form of surface or substrate to which a reagent solution is applied. Many of these kits are complex and include multiple tests for specific agents or families of agents. Color change detectors can detect nerve, blister, and blood agents.

Raman Spectroscopy is based upon the observation that when radiation is passed through a transparent medium, chemical species present in that medium scatter a portion of the radiation beam in different directions. The wavelength of a very small fraction of the radiation scattered differs from that of the incident beam. The difference between the scattered radiation and incident beam corresponds to wavelengths in the mid-infrared region. The degree of wavelength shift is dependent upon the chemical structure of the molecules causing the scattering. During irradiation, the spectrum of the scattered radiation is measured with a spectrometer. Raman spectroscopy appears not to be applicable for detecting CWA precursors and degradation products in soil samples but has applicability in air samples.

Mass Spectrometry (MS). A sample is introduced into the instrument, a charge is imparted to the molecules present in the sample, and the resultant ions are separated by the mass analyzer component. MS instruments are actually measuring the mass to charge ratio of the ions. A mass spectrum appears as a number of peaks on a graph. This technique only requires a few nanomoles of sample to obtain characteristic information regarding the structure and molecular weight of the analyte. Many mass spectrometers are specifically designed to detect various CW agents and have enormous applicability in detecting agents in most types of samples.

Gas Chromatography (GC) detectors are used to detect a variety of CW agents. Samples are subjected to a volatile solvent extraction. A small sample of the mixture is then injected through a rubber septum into a heated injection port that vaporizes the sample. The vaporized sample is then swept onto the column by the inert carrier gas and serves as the mobile phase. After passing through the column the solutes of interest generate a signal for a recording device to read. The detector is universal in nature in that it can respond to any change in the column effluent or only to solutes possessing a specific characteristic. Like mass spectroscopy, this method also offers high sensitivity and specificity in detecting CWA in many sample forms.

Fourier Transform Infrared (FTIR) Spectrometry. FTIR is a technique that can identify compounds that are separated by gas chromatography. After the separation of the compounds, the sample passes through a light pipe where an infrared (IR) beam is passed through it. The adsorption of the IR energy is monitored as the signal is continuously scanned. Scans are collected on each peak and the signals are then manipulated with a Fourier transform that enhances the signal to noise ratio of the spectra taken. FTIR detectors are used to detect a variety of CW agents.

MMST Equipment

The Metropolitan Medical Strike Teams (MMSTs) being organized and equipped by the Public Health Service (PHS) have purchased M8 and M9 detection paper, M256A1 Detection Kits, M18 Detection Kits, Draegar kits, portable surface-acoustical-wave (SAW) chemical agent detectors (SAW MiniCAD), and chemical agent monitors (CAM).

The M8 and M9 detection papers provide rapid (<1 minute), inexpensive tests for the presence of liquid mustard or nerve agents. Use of the paper is a screening test only, and results must be verified with more accurate methods of detection, particularly because of the paper's propensity to show false positive results for substances such as petroleum products and antifreeze. False positives are especially undesirable in a civilian context, where the mere rumor of "nerve gas" may cause hysteria.

M8 paper is supplied in the M256A1 Kit and the M18A2 Chemical Agent Detection Kit. M8 paper is a preliminary detection technique best suited for detection of liquid CWA on non-porous materials. M8 paper is tan in color and comes in a booklet containing twenty-five 2.5 inch × 4 inch perforated sheets. There are three sensitive indicator dyes suspended in the paper matrix. The paper is blotted on a suspected liquid agent and observed for a color change. V-type nerve agents turn the M8 paper dark green, G-type nerve agents turn it yellow, and blister agents turn it red.

M9 paper is an adhesive-backed, tape-like material designed to be worn on the outside of clothing or placed on vehicles, equipment, or supplies that may be exposed to liquid CWA droplets. The detector responds with a marked, contrasting color change, turning from the original green to red or pink when it comes in contact with a liquid CWA droplet.

The M256A1 kit includes enzyme-based detector "tickets," which change color to indicate low concentrations of cyanide, vesicant, and nerve agents in vapor form. The tests take approximately 15 minutes. Sensitivity is such that the tickets may provide a negative reading at concentrations below that immediately dangerous to life and health (IDLH) but still as much as 500 times greater than the acceptable exposure limit (AEL). Occupational Safety and Health Administration (OSHA) rules call for the use of maximum personal protection until concentrations can be shown to be less than 50 times the AEL. The IDLH is the maximum concentration of a contaminant to which a person could be exposed for 30 minutes without experiencing any escape-impairing or irreversible health effects. The AEL is a general term indicating a level of exposure that is unlikely to result in adverse health effects.

The M18 detection kit, like the M256A1 kit, is a military item. In fact it might be termed a chemical weapons Draeger tube kit—a colorimetric device for measuring the concentration of selected airborne chemicals. The M18 comes with disposable tubes for cyanide, phosgene, Lewisite, sulfur mustard, and nerve agents GA (tabun), GB (sarin), GD (soman), and VX. Tests for each take only 2 to 3 minutes but must be conducted serially.

The SAW MiniCAD is a commercially available pocket-sized instrument that can automatically monitor for trace levels of toxic vapors of both sulfur mustard and the G nerve agents with a high degree of specificity. The instrument is equipped with a vapor-sampling pump and a thermal concentrator to provide enriched vapor sample concentration to a pair of high-sensitivity coated SAW microsensors. All subsystems are designed to consume minimal amounts of power from onboard batteries. Optimal use of the SAW MiniCAD requires that a suitable compromise be made among the conflicting demands of response time, sensitivity, and power consumption. Maximum protection requires high sensitivity and a rapid response. The SAW MiniCAD is able to achieve a high sensitivity with an increased vapor sampling time. However, a faster response can be achieved at a lower sensitivity setting. Testing of the SAW MiniCAD has been performed with chemical warfare agents GD, GA, and HD. These tests were performed at a variety of concentrations and humidity levels. There were no significant effects noted due to the changes in the humidity levels for any of the chemical agents tested.

The CAM uses ion mobility spectrometry to provide a portable hand-held point detection instrument for monitoring nerve or vesicant agent vapors. It provides a graduated readout (low, medium, high). Response time is dependent on concentration but generally takes from 10 to 60 seconds. Minimum levels detectable are about 100 times the AEL for the nerve agents and about 50 times the AEL for vesicants. An obvious drawback to this relative insensitivity to low concentrations is an inability to fully check the efficacy of decontamination efforts, both in the field and subsequently at treatment facilities.

Few local governments or private medical facilities or organizations have invested in CWA detection equipment to date. This may change as the Army's Domestic Preparedness Program provides the training it has promised to 120 of the nation's largest metropolitan areas, but it seems likely that it is not simply information on availability but also the cost of these devices that has limited their procurement. For example, the CAM, a highly specific device designed to detect nerve and vesicant vapors only, costs almost $7,500. The equipment needs of early civilian responders to a domestic incident in which CWA may have been used are also different from those of military personnel in that the military has the advantage of intelligence information that enables the users of the equipment to predict a probable threat agent and the likely area of impact from the chemical agent. For first responders to a domestic terrorist incident, there are currently no such benefits of intelligence. Without such knowledge, first responders will be unlikely to use CWA-specific detection equipment immediately.

Much of today's technology has been developed into commercially available detection equipment, however, and this equipment should allow first responders, whether they be police, fire, Hazmat, or EMS units, to detect the presence or absence of CWA. This equipment is available, reasonably priced, and will detect a wide array of chemical agents. The M9 paper and the M256 kit are simple and inexpensive devices that enable responders to rapidly detect classical CW agents. The photo-ionization detector, the ion mobility detector, the surface acoustic wave detector, and the colorimetric tubes give medical personnel an ability to deal with a wider array of chemicals. As a market evolves for these items of detection equipment, modifications for the civilian community will be made to simplify their usage and the costs associated with their acquisition and maintenance should decrease.

Potential Advances

Current R&D in chemical agent detector technology is focused on increasing the speed and sensitivity of the instruments, while at the same time bringing down their size and cost. The vast majority of next-generation chemical detectors are based on the application of technology previously discussed. In some cases, the improvement will often be utilization of multiple technologies to simultaneously increase the sensitivity and specificity of the instrument. New CWA detector platforms with near-term successful development prospects include:

• Automatic Chemical Agent Detection Alarm (ACADA-M22). The M22 is an advanced, point-sampling, chemical agent alarm system employing ion mobility spectrometry. It is designed to detect standard nerve and vesicant agents.

• Improved Chemical Agent Point Detection System (IPDS). The IPDS also employs ion mobility spectrometry and is an improved version of a point detection system. In addition to G nerve agents and VX, the IPDS is designed to detect vesicant agent vapors. It is to be a shipboard instrument, and therefore will be much larger and need more power than portable IMS devices.

• Joint Chemical Agent Detector (JCAD). The JCAD will employ surface acoustic wave (SAW) technology to detect nerve and blister agents. It is designed to be lightweight and portable and will reduce false alarms. The JCAD will also allow detection of new forms of nerve agents.

• Joint CB Agent Water Monitor (JCBAWM). The JCBAWM will be a portable device to detect, identify, and quantify CB agents in water. It will allow the user to sample water and receive a digital readout of the contents. The technology to be employed in this monitor is still under review.

• Joint Service Lightweight Standoff Chemical Agent Detector (JSLSCAD). The JSLSCAD is a passive, infrared detection unit employing Fourier Transform Infrared (FTIR) Spectrometry. The device is designed to detect nerve and blister vapor clouds at a distance of up to 5 km.

• Shipboard Automatic Liquid Agent Detector (SALAD). Technologies to be used in the SALAD have recently been reviewed, but no decision has been made on the final selection. The instrument is designed to be an automated, externally mounted liquid agent detector capable of detecting G nerve agent and VX and vesicant chemical agents.

• The Special Operations Forces (SOF) Nonintrusive Detector and the Swept Frequency Acoustic Interferometry (SFAI) detector are portable, hand-held acoustic instruments developed specifically to enable rapid detection and identification of CW agents within munitions, railcars, ton containers, etc.

Various organizations are touting the advantages of new technology that offers the user the ability to determine the composition of potential CWA material within various containers without the need for direct sampling. Prototypes of the SFAI instrument employs a piezoelectric transducer that creates standing waves of sound in the container's contents. Software algorithms utilize the resonant peaks to extract the speed of sound through the contents, the density of the contents, and the attenuation of the returned sound signal. This type of instrument is highly sought after by explosive ordnance units, but offers little utility to medical personnel responding to a chemical release.

Tremendous efforts, primarily sponsored by the DoD, are under way to improve chemical agent detectors. The advances that will be of greatest benefit to the first responding medical teams will be increased portability, greater ease of use, and increased reliability of the detector technology.

Where the application of new CWA detection technologies could be of greatest potential benefit to the medical community is in fixed medical facilities and patient transport vehicles for monitoring air samples for low levels of CWA that may cause occupational hazards. Additionally, stand-off detection equipment can also assist medical planners in obtaining pre-incident intelligence so critical to providing the appropriate emergency response.

Medical personnel must rely on accurate and timely information provided by the earliest responders on the scene. If medical teams are expected to be the earliest responders to the scene of a mass casualty incident involving chemical agents, then they should be provided with reliable detection equipment as well as training on the use of the equipment. There should be continued support for the Public Health Service efforts to equip Metropolitan Medical Strike Teams with effective and currently available chemical agent detection equipment. These detectors are reliable, relatively inexpensive, and provide for the detection of all classical chemical agents that may be utilized in a domestic terrorist incident. Furthermore, efficient and cost-effective portable hand-held CWA detectors employing photo ionization detectors, surface acoustic wave microsensors, or ion mobility spectrometry should be readily available to all Hazmat units expected to respond to a potential CWA incident.

Information must flow along clear lines of communication and there must be standard procedures for relaying vital detection information. Rapid and sensitive assays for CWA could assist in limiting further exposures as well as providing verification and justification for initiation of appropriate therapy. Most importantly, medical response teams must be educated as to the nature and properties of CWA, be trained in recognizing the signs of CWA exposure, and be prepared to treat the symptoms caused by their toxicity. Detection and identification of the CWA is critical for legal and forensic purposes and for minimizing the transfer of contamination to unprotected personnel. However, detection and identification of the agent must not be the primary goal of the early medical response units; rather, it must be seen only as an aid to them in providing rapid and appropriate medical services to the victims of the incident.

CLINICAL LABORATORY ANALYSIS FOR EXPOSURE TO CHEMICAL WARFARE AGENTS

Rapid diagnosis of patients who have been exposed to a chemical agent will be important to saving lives and preventing further injury. The signs and symptoms of the patient will provide the most important information on which to base emergency treatment. Generally, instead of detecting an agent in the body, the clinician must look for some byproduct of the agent or a particular biochemical interaction within the patient that would indicate that an exposure to an agent has taken place. The specific biochemical interactions (such as, cholinesterase inhibition, thiodiglycol in urine) then lead the clinician to a determination of the likely agent. This section will examine methods for clinical analysis of four categories of agents: nerve agents, vesicating agents, respiratory agents, and cyanide.

Nerve Agents

Persons exposed to high concentrations of organophosphorus nerve agents usually develop signs and symptoms within a matter of minutes after exposure. Therefore, initial patient diagnoses and treatment are likely to be based on observations of signs and symptoms by the paramedic or other health care professional on the scene. Emergency medical personnel are, in any case, not equipped, trained, or encouraged to attempt clinical chemistry.

At the hospital level, treatment will continue to be guided by vital signs and clinical symptoms and monitored at this level by electrocardiogram, pulse oximetry, chest X-ray, arterial blood gas measurement, and other measures of physiologic status. Because nerve agents inhibit cholinesterase activity, laboratory tests estimating the level of this activity in red blood cells or plasma are sometimes used in estimating the degree of acute exposure. However, many hospitals cannot perform this test on site. Enzyme inhibition may only be loosely correlated with clinical signs and symptoms, and, because of high interindividual variability, only comparison between a baseline level of inhibition and the level just after exposure to a nerve agent will provide unambiguous evidence of a small or moderate exposure to nerve agents. A good example of this was reported in a recent study of 66 Japanese victims exposed to sarin (Masuda et al., 1995); patients exhibiting moderate symptoms of intoxication had serum values ranging from 300–750 IU/L. Normal serum cholinesterase activity ranges from 182 to 804 International units per liter (IU/L). These patients had red blood cell (RBC) ChE activity ranging between 0.3 and 2.0 IU as compared to 1.2–2.0 for patients not showing symptoms. Plasma ChE recovers in 30 to 40 days and RBC ChE recovers in 90 to 100 days after exposure to organophosphorus nerve agents (Grob et al., 1953).

Methods for Measuring Cholinesterase Inhibition

The standard methodology for determining blood ChE inhibition is based on the measurement of the enzymatic products derived from either acetylcholine or acetylthiocholine as substrates. The simplest and most convenient method is a colorimetric procedure (Ellman et al., 1961) in which thiocholine, the product of substrate acetylthiocholine hydrolysis, is detected by reacting with 5,5'-dithio-bis (2-nitrobenzoic acid) (DTNB). The time course of the reaction, monitored spectrophotometrically at 410 nm with and without various concentrations of the agent, is used to calculate the concentration of agent in the sample. A recently developed portable device utilizing this method, the Test-Mate™ OP Kit (EQM Research Incorporated, 2585 Montana Avenue, Cincinnati, OH 45211), provides a rapid, reasonably sensitive and reliable assay for ChE inhibition from potential OP exposure. The kit may prove suitable for use in a wide range of contingencies. It is recommended that the kit be used to establish ChE levels when exposure to nerve agents is suspected; it can be utilized at the site of an incident or in a hospital setting. This kit can determine RBC AChE and plasma butyrylcholinesterase (BChE) activities within minutes, requiring 10 µl of blood per determination.

Other procedures for determining serum ChE include liquid chromatography (Miller and Blank, 1991), an amperometric method using a hydrogen peroxide electrode, and the enzyme choline oxidase immobilized on a nylon net (Palleschi et al., 1990). Perhaps the most promising developments for screening kits for field use are immunochemical methods utilizing various murine monoclonal antibodies to acetylcholinesterase and ELISA (Novales-Li and Priddle, 1995).

Other Methods for Detection of Nerve Agent Exposure

A number of chemical methods have been used for the direct measurement of nerve agents and their metabolites in plasma and other body fluids. Among the most sensitive and reliable techniques are capillary gas chromatography (GC) (Bonierbale et al., 1997), gas chromatography-mass spectrometry, and gas chromatography-tandem mass spectrometry (Black et al., 1994). To date the most sensitive methods of retrospective detection of organophosphorus anticholinesterases are described by Polhuijs et al. (1997), who analyzed blood samples obtained from victims of the Tokyo subway incident. Fluoride ions are used to reactivate the inhibited enzyme, thus converting the OP moiety into the corresponding phosphofluoridate. It is suggested that this sensitive assay can be of benefit in biomonitoring of exposure for health surveillance, in cases of suspected use of nerve agents or pesticides, in medical treatment of OP intoxication, and in forensic cases against individuals suspected of handling anticholinesterases. Of practical utility are the GC/MS approaches of the U.S. Army (TB MED 296) for the determination of metabolites of sarin, cyclosarin (GF), and soman in urine. In animals exposed to the toxic organophosphorus nerve agents, substantial amounts of the parent compounds are hydrolyzed to their corresponding phosphonic acids (the rest is covalently bound to enzymes and tissue proteins) (Harris et al., 1964; Reynolds et al., 1985; Lenz et al., 1984, 1987). These methods are designed to detect these polar acids for verification of exposure. Urinary excretion of the metabolite is the primary elimination route for these three compounds. The major differences among them are primarily the extent and rate of excretion. Nearly total recoveries of the given doses for sarin and GF in metabolite form were obtained from urine, while soman was excreted at a slower rate with a recovery of only 62 percent. In animal studies, the acid metabolites can be detected in urine for 4 to 7 days post-exposure. The development of ELISA and monoclonal-antibody-based detection systems holds great promise in simplifying and hastening the detection of nerve agents in biological samples. Efforts have been made to develop immunoassays to chemical warfare agents (Lenz et al., 1992, 1997). Monoclonal antibodies were developed against a structural analog of soman, and, when employed in a competitive inhibition enzyme immunoassay, were capable of detecting the nerve agent at a level of 80 ng/ml.

Vesicating Agents

Methods for the detection of mustard have been described using GC/MS, based on the selective release by Edman degradation of the N-terminal valine adduct of hemoglobin with the agent (Fidder et al., 1996a; Noort et al., 1997) or by the determination of N7-(2-hydroxyethylthioethyl)-guanine, a novel urinary metabolite of sulfur mustard (Fidder et al., 1996b). The most abundant adduct, N1/N3-(2-hydroxyethylthioethyl)-L-histidine, is analyzed by LC-tandem MS, enabling the detection of exposure of human blood to 10 mM sulfur mustard *in vitro*. Verification of exposure to sulfur mustard in casualties of the Iran-Iraq conflict was conducted using these methods (Benschop et al., 1997). Gas chromatography-tandem mass spectrometry (GC-MS-MS) is also used to analyze urinary metabolites of sulfur mustard, derived from the beta-lyase pathway and from hydrolysis (Black et al., 1994). Another procedure utilizes GC-MS with ion chemical ionization to detect alkylated valine and histidine adducts of hemoglobin from casualties of sulfur mustard poisoning (Black et al., 1997a,b). Methods currently employed by the Theater Army Laboratory are those described in TB MED 296 (U.S. Army, 1995). In general, mustard cannot be simply assayed from urine because of its reactive nature. Thiodiglycol (TDG) is one of the *in vivo* degradation products of HD and can be used to confirm an exposure (Jakubowski et al., 1990; Davison et al., 1961; Roberts et al., 1963), although TDG is itself subject to chemical and enzymatic transformations. In this method, detection of TDG after derivatization with heptafluorobutyric anhydride is achieved by using a gas chromatograph coupled with a mass selective detector. The lowest quantifiable concentration is 5.0 ng/ml. Thiodipropanol is used as a stabilizer and octa-deuterated thiodiglycol as an internal standard. Again the advent of immunoassays for the detection of CW agents in body fluids holds the greatest promise in simplifying and reducing the time and cost to detect the presence of agents or their metabolites (Lenz et al., 1997).

Respiratory Agents

For sensitive and relatively specific detection of phosgene, a spectrophotometric GC-MS technique has been used that utilizes a reagent consisting of 0.4% of nitrobenzyl pyridine and 0.5% of sodium acetate in ethanol (Dangwal, 1994). The method is so sensitive as to detect phosgene at a concentration of 0.1 microgram/ml in the sampling solution with the coefficient of variation (CV) of 4.5%.

Cyanide

Four main laboratory findings are indicative of cyanide exposure: (1) an elevated blood cyanide concentration is the most definitive (most medical centers are unable to perform this measurement); (2) metabolic acidosis with a high concentration of lactic acid (lactic acidosis may result from a variety of conditions and is not in itself evidence of cyanide poisoning); (3) oxygen content of venous blood greater than normal (very difficult to measure and not specific for cyanide poisoning); and (4) presence in blood of cyanohemoglobin, which shows a characteristic absorption spectrum by UV spectrophotometry. As in the case of the nerve agents, however, the effects of cyanide (syncope, seizures, coma, respiratory arrest) occur so rapidly that treatment must begin long before laboratory findings are available.

Various methods have been used for CN^- detection in the blood (Feldstein et al., 1954; Lundquist et al., 1985). Most of them involve prolonged specimen preparation using diffusion or bubbling procedures, both of which require large blood volumes to achieve desired sensitivity. A sensitive and simple method for determining cyanide and its major metabolite, thiocyanate, in blood involved derivatization and determination by gas chromatography and mass spectrometry (Kage et al., 1996). The detection limits of cyanide and thiocyanate were 0.01 and 0.003 mmol/ml, respectively, while the gross recovery of both compounds was 80 percent. In another gas chromatographic procedure, cyanide was converted to cyanogen chloride by reaction with chloramine T and the product analyzed by electron-capture with a detection limit of 5 mg/L (Odoul et al., 1994). Cyanide determination in whole blood can be performed by spectrophotometry after using diffusion coupled with coloration by hydroxycobalamin in a Conway dish. The technique may be accelerated by the use of a heating sheet at 45°C. The method proved to be specific, sensitive, and fast, thus permitting measurements in emergency situations. A possible alternative to the GC-MS approach described above is an automated fluorometric measurement described in TB MED 296 (Groff et al., 1985). The CN⁻ assay methods provide direct measurement of plasma-free CN⁻ and the stabilization of total CN⁻ in blood. Samples for both plasma-free CN⁻ and total-blood CN⁻ are assayed directly without prior isolation of CN⁻ by a completely automated method requiring only 16 minutes from sampling to readout.

R&D NEEDS

R&D needs in chemical detection vary from simply evaluating what already exists to developing better procedures and methods for better detecting exposure levels. The ultimate goal of chemical detection is to rapidly and inexpensively detect and perhaps identify toxic substances that threaten to endanger both responders and victims. To this end, the committee has identified the following list of research needs:

4-1 Conduct a thorough evaluation of all industrial chemical detection equipment in the inventory of Hazmat and EMS units for its sensitivity and specificity for detecting CWA.

4-2 Continue research efforts to miniaturize and reduce the acquisition costs of GC/MS technology that would monitor the environment within fixed medical facilities and patient transport vehicles.

4-3 Develop standard operating procedures for communication of CWA detection information from first responders to Hazmat teams, emergency medical services, and fixed medical facilities.

4-4 Direct research efforts towards the development of simplified, rapid, and inexpensive methods of determining exposure to and level of intoxication from chemical agents in clinical samples. Give highest priority to research focusing on immunoassays for such detection, but research should also be conducted to determine the suitability of currently available portable instrumentation for rapid determination of cholinesterase inhibition by EMS units and in fixed medical facilities.

5

Recognizing Covert Exposure
in a Population

Most of the previous chapter assumes a scenario much like the explosives-based incidents that have been typical of terrorism to date—a sudden and highly localized event producing casualties almost immediately. This is a reasonable assumption for incidents involving most chemical agents, but it not likely to be accurate in the case of sulfur mustard, whose effects may not be obvious for several hours, and nearly all of the biological agents, whose effects are almost always delayed. This chapter will therefore focus on what we have previously termed scenario two, a covert attack with an agent producing signs and symptoms in those exposed only after an incubation period of several hours, days, or weeks, when the victims might be widely dispersed.

In these circumstances, effective medical response to many covert terrorist actions will be critically dependent not upon Hazmat teams, emergency medical technicians, and other "first responders," but upon the ability of individual clinicians, who may be widely scattered around a large metropolitan area, to identify, accurately diagnose, and effectively treat an uncommon disease. Education about the threat posed by bioterrorism, and about the diagnosis and treatment of possible agents deserves high priority, but the agents most frequently considered threats are rarely seen in U.S. cities and are therefore likely to remain quite far down any differential diagnosis. The identification of a single outbreak or a series of unusual disease presentations or deaths by the local or state public health department may therefore be the first clue that a cluster of disease may be related to the intentional release of a biological agent, unless the perpetrators reveal themselves. An analysis of the distribution and number of reported cases will provide important clues regarding the source of infection and can be used both to guide law enforcement and to help all physicians in the community make a rapid and accurate diagnosis of new cases and begin optimal treatment without delay. A very large and efficient attack may truncate the dispersion of cases in space and time (even in this case, victims will not be affected immediately, as they would be with chemical agents), making effective intervention difficult even as it makes it more obvious that the cause is a deliberate release. Rapid and accurate epidemiologic investigation will nevertheless be a key factor in minimizing suffering and loss of life in bioterrorist incidents. Surveillance systems for collecting reports of such cases and appropriately trained staff to monitor for disease outbreaks are the foundation of public health epidemiology. Yet over the last few decades there has been severe erosion in the capacity of public health departments to conduct disease surveillance and epidemiologic investigation.

SURVEILLANCE AND INVESTIGATION
OF BIOLOGICAL AGENTS

A rapid evaluation by public health epidemiologists is absolutely critical. Delays in determining the scope and magnitude of the exposure may result in illness and deaths that might have been avoided if a rapid response, based on accurate and timely surveillance data, was made.

Surveillance systems can be passive or active. The large majority of surveillance systems in place at local, state, and federal levels is passive. These rely on systems of disease reporting from health providers and are notorious for their poor sensitivity, lack of timeliness, and minimal coverage. They are inexpensive to implement, but the quality of information is greatly limited, and most are not well suited to the needs of modern disease surveillance, including that needed in the case of a biologic terrorist event. The single exception appears to be electronic laboratory reporting (described below), which can give useful and timely data for epidemiologic purposes.

The Centers for Disease Control and Prevention (CDC) oversees a large number of passive infectious disease surveillance systems. These systems are based on voluntary collaboration with state and local health departments, which in turn depend on physician-initiated reports of specific diseases or information from state health laboratories regarding bacterial or viral isolates. The best known system is the National Notifiable Disease Surveillance System, which the CDC describes as the backbone of collaborative reporting procedures involving clinicians, state, and local health departments, and the CDC. Clinicians, hospitals, and laboratories in each of the 50 states, the District of Columbia, and the territories are required (by their own laws) to report cases involving any of a list of approximately 50 diseases. The list is compiled and periodically revised by a collaboration of state and CDC epidemiologists (federal agencies cannot legally dictate to states which diseases should be reported). It currently includes several of the diseases this report focuses on as possible bioterrorism agents: anthrax, botulism, brucellosis, plague, and eastern and western equine encephalitis.

Each state and territory has its own list of reportable conditions, most of which overlap with the national list. Certain other medical conditions likely to be caused by bioterrorists may be reportable conditions in selected states, but this is not consistently true across the nation.

The reliability of passive surveillance systems is often quite low, especially if a physician or hospital fails to make the initial report or does not do so in a timely manner. While many states may have legal penalties that can be brought to bear against a provider who does not report, such penalties are almost never imposed. Neither are there any real incentives to report. In the case of an illness due to an exotic biological agent, reporting to proper public health authorities may be more likely to occur than with an illness caused by a common pathogen. If a terrorist uses a common pathogen, it may be difficult to determine the mode of transmission, considering the normal background disease incidence.

Little or no federal funding is provided to state and local health departments to support the surveillance of general communicable diseases. The ability of the states to support infectious disease surveillance has declined in recent years. A 1993 survey indicated that 12 states had no professional position dedicated to surveillance of foodborne and waterborne diseases (Osterholm et al., 1996). Although most states do have some infectious disease surveillance capacity, it is most often supported by categorical (i.e., disease-specific) funds, for example, Acquired Immune Deficiency Syndrome, vaccine-preventable diseases, sexually transmitted diseases. These funds generally cannot be used to support noncategorical communicable disease activities.

CDC has begun an Emerging Infections Program (EIP), under which grants are awarded to state or local health departments for improving epidemiological and laboratory capability. One of the basic benefits the EIP provides for state-based communicable disease surveillance is more trained professionals to follow through with investigations that might not have been initiated because of limited resources. This is one effort designed to ensure there are enough trained public health epidemiologists to maintain a working surveillance system, follow up on cases obtained from that system, and conduct the necessary investigation to develop evidence for causation.

In seven states supported by EIP, active population-based surveillance for selected diseases (foodborne diseases, opportunistic infections in inner-city HIV patients, community-acquired pneumonias, febrile and diarrheal illnesses in migrant farm workers, and unexplained deaths in young adults) is under way. Expansion of these or similar efforts to all fifty states, if accompanied by an aggressive telecommunications effort to make the resulting data widely and easily available, would be a good start towards remedying the domestic surveillance shortfalls identified by a previous IOM report (Lederberg et al., 1992). Although certainly not adequate in itself, this remedy could serve as the model for improving the eroded public health surveillance and investigation infrastructure necessary for preparing the country to adequately respond to a biological agent release.

Active surveillance, which requires staff to actively search for and identify new cases, provides more timely and accurate information than the commonly used passive systems. To conduct active surveillance, state and local health departments must have sufficient numbers of adequately trained epidemiologists. These scientists collect, compile, analyze, and interpret the epidemiologic data to determine the source of the infectious agent. Additionally, capacity is required to conduct the field investigations that are dictated by the surveillance data or by case reports.

A national effort to improve active reporting is the expansion of the Sentinel Surveillance Networks, supported by the EIP program. In this effort, selected reporting relationships are established with medical specialty groups like the Infectious Disease Society of America, the International Society of Travel Medicine, and a group of about 100 emergency departments called the Emerging Infectious Disease Network. These specialists are likely to treat persons with infectious diseases and are periodically contacted about any specific or unusual conditions.

Further research into the potential of electronic disease reporting from (and to) physician's offices is also warranted. Large medical practices, public health clinics, and managed care settings often have patient medical records in electronic form, or other electronic documentation that may identify a case of a reportable (or potential biological terrorism-related) illness. It would be helpful to determine how to develop the capacity to have certain records and information automatically sent to proper health authorities. The legal authority for release of certain information already exists in most state health codes and some local health ordinances. Issues of confidentiality and security must be addressed at the same time as the technological constraints are tackled. An important group of professionals that should not be overlooked in such a surveillance effort is the veterinary medicine community. They are often familiar with a number of the biological agents (anthrax, plague, brucellosis, tularemia, and the equine encephalitides are all far more common in animals than in the U.S. population), and their incidence and prevalence in local livestock and wildlife.

The CDC is currently examining the development of an emergency department (ED)-based surveillance system called Data Elements for Emergency Department Systems (DEEDS). The DEEDS system is designed to standardize electronic emergency department reporting across clinical systems of care. The system could be the core of a reporting process which identifies diseases of public health importance on a daily basis. Systems such as this could be strengthened to provide real-time surveillance systems of communities at risk. Such a system would not only identify established diagnoses but could also report the actual laboratory request for certain diagnostic assays. These tests would include those done to look for unusual diseases with potential for biological terrorism (i.e., a culture for anthrax). Other systems that track the utilization of specific antibiotics or vaccines (i.e., botulism) could also serve as an early warning system.

The DoD Global Emerging Infections Surveillance and Response system (DoD-GEIS) is designed to conduct antibiotic resistance surveillance at the six DoD tropical medical research units and coordinate emerging infections surveillance for the three services. The U.S. Air Force Global Surveillance Program is an attempt to integrate several surveillance efforts in a way that tracks emerging diseases. Systems such as this could be utilized to track diseases of concern for biological terrorism.

Internationally, projects such as the Canadian Bacterial Disease Network (CBDN) link university and government laboratories in an effort to track diseases of importance. Health Canada, Canada's federal health department, is developing a system to scan the Internet for evidence of new disease outbreaks, utilizing a new search engine technology designed to constantly search a large number of predesignated sites, popular as well as scientific. The system, called the Global Public Health Intelligence Network (GPHIN), will identify disease reports and link them to the World Health Organization's information system.

Numerous other surveillance systems track disease transmission and antimicrobial resistance both nationally and internationally (Appendix B), but there are few meaningful links among them. Mechanisms to integrate and link these systems should be developed.

Detecting and characterizing an outbreak caused by a covert release of a biological agent can be difficult, or it may be startlingly obvious. A reported case of anthrax in an area of the country where anthrax is never reported or in an individual with no obvious risk factors for the disease would raise the suspicions of the public health epidemiologist. Although intentional infection would not necessarily be the first explanation investigated, a process of elimination or additional case reports would eventually lead to serious consideration of this possibility. The time it takes to reach this point is the rate-limiting step in societal response. It is, then, a critical infrastructure resource and expertise problem of national importance. Without a sufficient number of adequately trained epidemiologists at the local and state level, there may be significant delays in identification and response. In the case of a biological event, lost time may quickly translate into lost lives.

Formal training of epidemiologists occurs in schools of public health, medical and veterinary schools, and in on-the-job-training in health departments around the nation. The number of epidemiologists who are prepared for field public health work is limited. The applied public health sector competes poorly with academia and industry for new epidemiology graduates. In addition to those trained at medical schools and schools of public health, the CDC trains a cadre of Epidemic Intelligence Service officers (EIS), who are available to assist state and local epidemiological response. The EIS was created during the Korean War in response to fears about biological weapons and the perception that state and local public health resources were inadequate to deal with disease outbreaks (Langmuir and Andrews, 1952). Now, nearly 50 years later, facing a threat from these same weapons in the form of biological terrorism, our nation still finds itself understaffed and underprepared.

Current public health epidemiology staff often lack access to authoritative guidelines and other information regarding treatment and control measures. In medium to small jurisdictions, the person serving as the epidemiologist may have little formal training in the concepts and methods of field epidemiology and surveillance and have no background in biological or chemical threats. Most large states and cities employ trained epidemiologists, but few of those individuals have any sophisticated knowledge about biological or chemical weapons. There are national experts who have a significant amount of knowledge regarding biological agents and the consequences of their release, but there is little interaction between these experts and the front-line health department epidemiologists.

No hard data exist on the current knowledge of state and local health departments regarding chemical and biological terrorism. From that information, specific educational and training efforts could be developed and implemented for all state and local health departments. These solutions might include the development of training meetings, Internet-based training, video conferences, or other information exchange technology aimed at the education of local and state health department staff. Syllabi might include not only the medical perspectives of the diseases of interest, but also information regarding modes of transmission, laboratory considerations, working with emergency responders, educating physicians and other health care providers about these conditions and reporting them to the proper health agency, psychological aspects, and the special considerations of intentional release of these agents. Additionally, since health departments will be requested to disseminate information about the agents, the circumstances of the release, and control measures, information packages for public use will need to be developed. The Centers for Disease Control and Prevention and national organizations, such as the Council of State and Territorial Epidemiologists (CSTE), the National Association of County and City Health Officials (NACCHO), and others may be well suited to develop and facilitate this training.

In the last decade, advances in information technology have greatly accelerated the speed at which business and commercial transactions and information exchange occurs. However, these advances have, in large part, not reached local, and in some cases, state health agencies. The capacity of state and local health departments to communicate electronically with each other is severely limited. Fewer than 50 percent of local health departments have any capacity for Internet connectivity (electronic mail) (National Association of County and City Health Officials, 1997).

One current system of communicating news and information about worldwide emerging diseases utilizes the Internet and electronic mail capabilities. A project of the Federation of American Scientists, ProMED (Program for Monitoring Emerging Diseases) was established to provide communication among sentinel stations around the world capable of detecting unusual outbreaks of infectious diseases or toxic exposures, including those that might result from a biological attack. Scientists from around the world, including many national and global infectious disease experts, participate in a daily exchange of information regarding infectious disease cases and outbreaks. Currently there are more than 10,000 participants in this communication system, which provides a forum for discussing disease occurrence. While ProMED may not be directly useful in emergency notification of authorities of an unusual, potential biological agent release, it could serve to quickly bring experienced scientists together electronically for discussion of the situation. This would improve the response capabilities of public health departments. Additional research into the use of ProMED or other Internet-based information sources (e.g., Outbreak, and Communicable Disease Prevention and Control) appears warranted.

LABORATORY CAPACITY AND SURVEILLANCE

The variable and often substantial delay between exposure to a biological agent and the onset of clinical signs and symptoms, as well as the possibility of person-to-person transmission, makes rapid and accurate diagnosis important, even if treatment of the earliest patients cannot be guided by laboratory findings. Protection of healthcare workers and treatment of delayed victims of the attack and secondary victims infected by contact with an early victim will be much enhanced if exposure can be confirmed and treatment started prior to symptom onset.

In each case of suspected exposure, appropriate diagnostic samples from blood, serum, stool, saliva, or urine are needed for laboratory identification of the specific agent. Franz et al. (1997) list the following diagnostic assays: (1) Gram's stain for anthrax and plague; (2) serology, including enzyme-linked immunosorbent assay (ELISA), agglutination, immunofluorescent assay (IFA), hemagglutination inhibition, and antibody (AB) ELISA for anthrax, plague, Q fever, tularemia, viral encephalitis, viral hemorrhagic fevers, botulinum toxin, and staphylococcal enterotoxin B; (3) culturing for brucellosis, plague, and tularemia; (4) Wright-Giemsa stain for plague; (5) virus isolation for smallpox, viral encephalitis, and viral hemorrhagic fevers; (6) electron microscopy for viral hemorrhagic fevers; and (7) polymerase chain reaction (PCR) for identifying the genetic material of smallpox and hemorrhagic fever viruses. These assays may take anywhere from 2 hours to 30 days to complete, and, in the case of smallpox and the hemorrhagic fevers, demand Biosafety Level 4 procedures (e.g., controlled-access laboratory, change to laboratory coveralls and shower on exit, work conducted in fully enclosed, separately ventilated biological safety cabinet [CDC and the National Institutes of Health, 1993]). Even research facilities with this level of protection are not common.

Although many hospitals and commercial laboratories have the necessary equipment and expertise to perform these and similar assays, these diseases are extremely rare in the United States, and so these laboratories rarely perform these assays. Most laboratories will thus not be prepared to immediately conduct the specific analytical test needed to identify the agent, even when the attending physician is astute enough to ask for the test. Veterinary diagnostic laboratories may be more likely sources for a rapid confirmatory assay, since many of the biological agents are significant sources of disease in domestic animals (nonhuman species may even be the first victims of a biological attack, unintentionally or as part of an effort to avoid discovery). Veterinarians and veterinary laboratory workers, moreover, are likely to have been vaccinated against many zoonotic diseases, and used to working with these agents. In no case, however, will a test be done unless a suspicious physician requests it. A more likely scenario is a round of tests for more common pathogens, followed by or concurrent with some symptom-based treatment. Continuing deterioration of the patient's condition or an unusually large number of affected patients may then lead to involvement of a state health department laboratory and state epidemiologist.

Although the capabilities of these laboratories vary widely among the states, all have working relationships with the CDC and can call upon CDC and other federal and university laboratories for help. Because these organisms and diseases are seldom seen in the United States, there are few experts in these diseases even at the CDC (additional evidence of resource erosion), and the CDC may need to call upon USAMRIID if one of these diseases is suspected. Although this channeling of samples from the initial round of victims to a single expert organization will help in identifying an outbreak, ensure that medical and laboratory personnel are protected, and facilitate rapid diagnosis of those delayed or secondarily infected patients, the process by its very nature is quite slow and will provide very limited benefit to the first victims. Research into the utility of developing a network of regional laboratories (including veterinary laboratories) capable of rapid diagnostic testing may be useful. Ideally, such a network would involve strengthening diagnostic expertise and laboratory capability at all major medical centers, but given the expected frequency of cases involving biological warfare agents, a more realistic goal might be a regional approach based on state and local public health laboratories.

Knowledge regarding the laboratory recovery of a pathogen or positive diagnostic assay is a key element that public health epidemiologists will need in their investigation. While, in general, an epidemiologic investigation can be initiated without laboratory confirmation, final confirmation of the exact nature of the pathogen will be needed. In some cases, the laboratory report may serve as the initial notification that there is a human illness associated with a microorganism that is known to be a potential terrorist agent. Such a report does not make a diagnosis, nor does it suggest how the pathogen was acquired by the patient. However, it could serve as an early notification system for public health, thereby improving the chances of responding quickly to additional cases and saving lives.

Currently, biological samples are taken from the patient and delivered to the laboratory for analysis. After identification through culture or diagnostic assay, the results are sent back to the treating facility for use in medical management of the patient. Most communicable disease surveillance systems rely on telephonic or weekly written reports for subsequent transmission of this information to local or state health officials. Even if a suspect atypical pathogen is identified and interested health care staff make special efforts to report, the time lag before the health department knows of the culture may exceed 3 days.

In most large laboratories, tracking of specimens and results is done electronically via computer. Recent efforts, spearheaded by the National Center for Infectious Diseases, at the CDC, and the CSTE have focused on the possibility of electronic reporting of laboratory results. This beginning effort has demonstrated that many large laboratories are capable of downloading assay findings in formats that are usable by state health officials.

Originally designed to reduce the workload of reporting results to over 50 jurisdictions, electronic laboratory reporting may serve as an important step towards getting important information into the hands of the epidemiologist quickly. Additional research towards full development, and then national implementation of electronic laboratory reporting will improve the public health response to a biological release. These solutions will also strengthen the national disease surveillance infrastructure in general, a glaring need detailed in the IOM report *Emerging Infections: Microbial Threats to Health* (Lederberg et al., 1992).

An important new laboratory development is the ability to sequence different parts of microbial genomes. By identifying distinct features of different genes, it is possible to identify not only microbes of interest but specific strains and thus more precisely track infectious disease outbreaks. This "fingerprinting" technique will be useful as a sentinel indicator that a new strain has entered a community and in distinguishing natural from intentional releases by identifying microbial or viral strains foreign to the normal community flora or by matching new outbreak pathogens with pathogen strains from suspect terrorist groups or intelligence sources. Such a library of genetic fingerprints would also have enormous value not only in the effort to track ongoing outbreaks, but also in predicting antigenic shifts for vaccine production..

The beginnings of a national network of state health laboratories utilizing this type of fingerprint technology and sharing fingerprint libraries is under way. Known as PulseNet, this system, officially initiated in 1998, links the CDC, U.S. Department of Agriculture, and the U.S. Food and Drug Administration to a network of state laboratories using pulsed-field gel electrophoresis to look for characteristic DNA patterns of organisms implicated in foodborne infections The system is currently capable of identifying patterns of *E. coli 0157:H7* and tracking them across 12 states. Expansion of this system to other states and additional pathogens is planned and could form the backbone of the national network suggested above. Including more states is relatively straightforward; including more pathogens is far more involved, and probably depends not only on identification of stable and accessible DNA and RNA sequences that are unique to particular pathogens or classes of pathogens, but also on simpler, faster, user-friendly microbiological assays. Chapter 6 will address the possibilities of such improvements and their application to both patient diagnostics and environmental monitoring and testing.

CHEMICAL/TOXIN SURVEILLANCE

Although the rapid onset of symptoms in a large number of victims may make it obvious when a chemical attack has taken place, poison control centers (PCC) can serve as the basis for a surveillance system for recognizing covert or multiple-site poisonings with chemical agents or biological toxins. The PCC is also the logical place to turn in that scenario for advice on treatment.

There are no studies defining the theoretical or actual skills of poison information specialists and physician staff with regard to chemical or biological warfare agents, but these individuals are well prepared to provide advice to emergency personnel in the field and at hospitals on the basis of signs and symptoms. The PCC may also serve as a coordination point, even when the "incident" is confined to a narrowly circumscribed locale, for victims may be dispersed to a number of different medical facilities. The Tokyo incident involved release of sarin in five trains on three different subway lines; 278 medical facilities received patients in the following 48 hours (Sidell, 1996). The value of a medical information coordinating center in such a situation cannot be overestimated. Unfortunately, poison control centers have also faced years of declining resources, with many centers performing few of the essential tasks to recognize, much less contain, a developing epidemic. This trend must be stopped or the nation may face a shortage of medical toxicology specialists who can provide critical treatment information to treating physicians. Additionally, linkages to the public health system do not exist in many states and should be encouraged.

At the national level, the Agency for Toxic Substances and Disease Registry (ATSDR), instituted a hazardous substances emergency events surveillance (HSEES) system in 1990 (U.S. Department of Health and Human Services, 1993, 1994b, 1995a). State health departments in selected states collect and transmit information on the circumstance and health outcomes surrounding hazardous materials releases. The information is generally provided well after the event. The information in HSEES is made publicly available for use in locating, training, and equipping Hazmat teams, first responders, and employees as well as guiding follow-up epidemiology. However, the time constraints associated with an incident of chemical terrorism will not permit a significant real-time surveillance role for the HSEES.

AIDS FOR CLINICAL DIAGNOSIS BASED ON SIGNS AND SYMPTOMS

Emergency medical personnel, both at the scene of a hazardous materials incident and at hospital emergency departments, have a wide selection of reference materials to call upon for guidance in patient management. These include traditional textbooks, handbooks such as the three-volume set of medical management guidelines prepared by the Agency for Toxic Substances and Disease Registry (U.S. Department of

Health and Human Services, 1994a), paper, CD-ROM (Poisindex, Drugdex, Emergindex), or on-line Material Safety Data Sheets. Some of these sources provide information on nerve and mustard agents, but most of these resources are organized by chemical rather than by symptom complex. That is, given some independent knowledge of the identity of the hazardous substance, one can readily ascertain the likely effects and appropriate treatment. Deducing the substance from the effects is far more difficult. Poison control centers are routinely faced with this problem and are a good source of help.

The Washington, D.C. MMST has addressed this difficulty by incorporating a symptom checklist tool called the "NBC Indicator Matrix" into their training (Defense Protective Service, 1996). First developed by the Defense Protective Service, which provides security at the Pentagon, it is a paper-and-pencil checklist of symptoms. A system for scoring and processing the results leads to a suggested agent or agents. For Hazmat incidents at the Pentagon, and by definition, for incidents which lead to a request for help from the MMST, first responders are highly likely to turn to such a tool. At other locales, they may need some reason to suspect terrorism to consider using the matrix. The matrix may have some utility even in its present form, especially in cases in which exposures are mild. Difficulties seem likely when victims are critically ill or have pre-existing illnesses, or in incidents involving more than a single agent.

With the exception of some of the toxins (botulinum, SEB, T-2 mycotoxin), and possibly the hemorrhagic fevers, the initial signs and symptoms produced by the biological agents considered here are nonspecific—fever, chills, fatigue, headache, muscle or joint pain, a cough or chest pain. Blood in the excreta or petechiae (pinpoint-sized, hemorrhagic spots in the skin) may lead an astute clinician to consider a hemorrhagic fever, but few U.S. practitioners are likely to recognize the other diseases associated with biological weapons on the basis of signs and symptoms alone. Correct diagnosis will almost certainly depend on perception of an unusual epidemiologic picture by public health epidemiologists. This is an area where pre-incident intelligence could have a major impact in reducing the number of casualties.

The development of interactive computer-based diagnostic systems that enhance the potential for early recognition and analysis of the unique aspects of rare diseases, including the manifestations of disorders produced by biological and chemical agents, would be a substantial advancement. An integrated system that utilizes natural disease rates and clinical probabilities, based upon signs and symptoms, and laboratory findings could enhance an early warning system prior to a clinician's decision on a particular diagnosis or disease. One model, the Global Infectious Disease and Epidemiology Network (GIDEON), uses a Bayesian matrix for compatible diagnoses. This particular system, which focuses on unusual diseases, is limited by the weight it places on country of disease origin and requires a database that might not be initially available.

An improved system with refinements in diagnostic decisionmaking that offers the untrained or inexperienced clinician assistance in considering a chemical or biological exposure would be of great value. For any system this would necessitate a complex, multiple search mechanism that includes early signs and symptoms of atypical disorders caused by biological or chemical terrorists agents.

R&D NEEDS

The committee recognizes that the first of the following recommendations is a recommendation, not for research itself, but for the prerequisites for productive R&D. The committee strongly believes that no research effort, no matter how important or sophisticated, will be productive until the nation rebuilds the public health infrastructure to a level at which the results of appropriate research can be properly applied. This infrastructure improvement would have enormous value to the average citizen on a day-to-day basis and would generate significant health benefits beyond readiness for terrorist events.

5-1 Immediately undertake improvements in CDC, state, and local disease and exposure surveillance and epidemiologic investigation infrastructure, and support them on a long-term basis. These improvements must focus on communicable disease epidemiology and laboratory programs and on poison control centers.

5-2 Evaluate the current educational/training needs of state and local health departments regarding all aspects of a biological or chemical terrorist incident. Develop and put in place programs and materials based on the research findings and aimed at preparing these departments and their health care partners to adequately identify and respond to such an incident.

5-3 Conduct research on new, faster, and more complete methods of electronic disease reporting to enhance surveillance at all levels, including the health care provider, local, state, national, and global surveillance levels. Such research should include evaluating the benefits of utilizing Internet and electronic mail technologies to improve reporting and access to expertise concerning biological or chemical weapons before, and during a release.

5-4 Enhance research efforts to develop nucleic acid fingerprinting techniques capable of tracking microbes likely to be used by terrorists. A library of these fingerprints, and the laboratory techniques to develop and use them should be available to a network of cooperating regional laboratories.

5-5 Conduct research into the development of symptom-based, automated decision aids that would assist clinicians in the early consideration and identification of unusual diseases related to biological and chemical terrorism.

6

Detection and Measurement of Biological Agents

The previous chapter was devoted to an analysis of what the committee feels is the most probable course of events in a terrorist attack involving a biological agent—a covert attack that, after a period of hours to weeks, will result in victims widely distributed in time and location. Because the biological agents being discussed in this report do not immediately produce effects, the first indication of an attack with a biological agent may be the recognition of an unusual distribution or number of cases of disease, long after the initial aerosol or solution has been dispersed or degraded. An important part of this detective work is laboratory analysis of clinical samples, most often blood from a sick patient. The previous chapter alluded to the possibility of new developments in such diagnostic testing that might significantly decrease the time needed to arrive at a definitive diagnosis. The present chapter examines those developments in more detail.

The chapter also examines the application and utility of these developments in the detection of biological agents in the environment. There will be no fire and rescue teams responding to a 911 call in an incident involving covert release of a biological agent, and thus little use for the sort of rapid detection devices that are so important in responding to chemical releases. Public health surveillance systems and the rapid analysis of information from those systems may in time provide an indication of when and where the biological agent was released, but unless there is a continuing source of agent, testing the release site at that point will probably be useful for forensic purposes only (testing may also be helpful in guiding clean-up after an attack with spore-forming agents like anthrax that can survive in the environment for years). This is far different from the battlefield scenario of military units facing an enemy with an arsenal of identified biological weapons. Monitoring the environment for those agents and providing these at-risk troops with the means to rapidly identify contaminated air, water, food, and equipment would literally be vitally important. To the extent that similar high-risk situations can be identified in the civilian environment (the President's State-of-the-Union Address? The Superbowl? A soggy package labeled "anthrax"?), there may be a civilian need for monitoring and detector technology as well. For this reason, although the committee does not believe these situations will be frequent enough to merit a major investment for civilian use, the chapter concludes by summarizing current R&D efforts on environmental detection by military and other laboratories.

DETECTION OF BIOLOGICAL AGENTS IN CLINICAL SAMPLES (PATIENT DIAGNOSTICS)

The classical approach to microbial detection involves the use of differential metabolic assays (monitored colormetrically) to determine species type in the case of most bacteria, or the use of cell culture and electron microscopy to diagnose viruses and some bacteria that are intracellular parasites. Samples taken from the environment, such as soil and water, and most clinical samples must be cultured in order to obtain sufficient numbers of various cell types for reliable identification. The time required for microbial outgrowth is typically 4–48 hrs (or two weeks for certain cases, such as *Mycobacterium tuberculosis*). Furthermore, bacterial culture suffers from an inherent drawback: cells that are viable may not be culturable, because they possess unanticipated nutritional requirements as a result of genetic mutation. The following few pages lay out some general approaches being taken to eliminate these drawbacks of the traditional methods and provide some examples of efforts to apply them to detection of potential biological weapons. Biodetection is a very large and active field which merits a study all by itself, and for that reason the rest of the chapter is deliberately confined to technologies and research that has focused on the agents of central concern to this report. The interested reader is referred to any or all of the following general reviews: Turner et al. (1987), Janata (1989), Wolfbeis (1991), Taylor and Schultz (1995), Van Emon et al. (1996), Rogers et al. (1995), Kress-Rogers (1997). Boyle and Laughlin (1995) provide a history of the U.S. military biodetection program, and Boiarski et al. (1995) described a large number of biodetection technologies being explored by the U.S. military at that time.

In summarizing the current review, it is convenient to consider detection of biological agents as a two-stage process involving: (1) a probe, and (2) a transducer. Probe technology deals with how the assay or detection device recognizes the particular target microbe. Transducer technology deals with how the assay or detection device communicates the activity of the probe to the observer. Together, probe and transduction systems determine specificity, sensitivity, and time required to make an identification.

Probe Technologies

Probe technologies include those based on: nucleic acids, antibody/antigen binding, and ligand/receptor interactions.

Nucleic acid-based probes capitalize on the extreme selectivity of DNA and RNA recognition. Nucleic acid probes, engineered single strands of RNA or DNA, bind specifically to strands of complementary nucleic acids from pathogens. These probes and their binding can be detected directly or by tagging with an easily detected molecule that provides a signal. The design of the probe can be highly specific if there is a good fit to a pathogen-unique region of the target nucleic acid, or it can provide more generic identification if there is a fit with a region of nucleic acids conserved among several related pathogens. The sensitivity of these hybridization assays for bacteria is between 1,000 and 10,000 colony-forming units; improved sensitivity is an important area of research. Since the reaction is in real time, the time-consuming part of the method relates to sample preparation and the time required to detect the signal.

The main advantages of nucleic acid-based methods are universality (all living organisms have DNA and/or RNA), specificity (every type of organism has some unique sections of DNA or RNA), sensitivity (with amplification, very small amounts can be detected), adaptability (base sequences common to several microbes, or even a whole class of microbes, can be used as probes), and multiplex capabilities for a host of different microbes (a sample can be probed for many different sequences simultaneously). Disadvantages of this technology include difficulty in isolation and "clean-up" of DNA samples, degradation of the nucleic acid probes, and interference from related sequences or products. These are important obstacles to be overcome, even after specific and accessible target sequences are identified and probes constructed.

Some commercial products are already available for applications unrelated to biological weapons. Parke-Davis, for example, markets an RNA-based device to study HIV RNA: protein interactions. A dozen or more biotechnology companies are pursuing production and a variety of applications of "DNA chips," microarrays of 100 to 100,000 or more DNA or oligonucleotide probes immobilized on glass or nylon substrates (Marshall and Hodgson, 1998; Ramsey, 1998). Santa Clara-based Affymetrix, for example, has developed a dime-sized GeneChip™ using arrays of 100,000 or more fluorescence-tagged hybridization probes and scanning confocal optical readout to search for mutations of genes known to be involved in specific human diseases. The readout instrumentation is expensive and the chips themselves have a shelf life of only a few months, but the speed and thoroughness of the search may have appeal for pharmaceutical and biotechnology companies. Roche Molecular Systems recently announced its intention to purchase GeneChip arrays for use in planned diagnostic kits for application to HIV drug resistance and cancer staging, and a collaboration among Affymetrix, Lawrence Livermore National Laboratory, and the U.S. Army Medical Research Institute for Infectious Diseases (USAMRIID) is attempting to adapt the technique to detect biological weapons as characteristic sequences are identified. A large array incorporating many more common pathogens as well might encourage everyday use in large medical labs and eliminate the bottleneck to rapid diagnosis identified in the previous chapter—the need for a suspicious clinician to order an assay for a very rare disease.

A similar but smaller microarray of gel-immobilized, fluorescence-labeled nucleic acids is being developed by Argonne National Laboratory (Yershov et al., 1996). One application seeks to develop a "bacillus microchip" that will detect *B. anthracis*, indicate whether it is alive or dead (DNA matches, but no RNA matches), and distinguish it from other related bacteria, such as *B. thuringiensis*, *B. subtilis*, and *B. cereus* (Mirzabekov, 1998).

A second application of the Argonne/3M array takes advantage of that latitude by employing RNA probes from the highly conserved 16S ribosome to provide a tentative taxonomic assignment to unknown bacterial pathogens, including novel or bioengineered organisms (Risatti et al., 1994; Stahl, 1998). This strategy will not work for all bioengineered organisms—identification by taxonomic markers must be supplemented by identification based on markers of pathology, however, if we are to successfully cope with harmless microbes provided with genes from pathogenic organisms.

Antibody-based probes (immunosensors) offer another highly specific probe technology, since antibodies recognize very specific sites or cellular components (epitopes). Antibodies specific for any microbe can be made if the microbe can be obtained in pure culture. These must be screened for binding characteristics, that is, binding affinity, on- and off-rates, and epitope recognized. The production of monoclonal antibodies requires significantly more time and effort in the development of hybridoma cell lines with appropriate characteristics. It is, therefore, desirable to provide for breaking the antibody-antigen bond after a positive test and reusing the antibody in additional tests. The binding of the target (antigen) to the antibody can be monitored directly with a transduction method, such as luminescence or electrochemical signal, or can be monitored in a sandwich assay in which a second antibody labeled with a fluorescent dye binds to another epitope on the captured cell or to the probe antibody. Indirect methods monitor the bound epitope by its competition with a standard epitope labeled with a fluorescent dye. While this indirect format is more sensitive, the antibody must bind very strongly to the antigen target.

Fluorescence-based fiber optic immunosensors have demonstrated the detection of 10^4 microbial cells/ml, and immunoelectrochemical sensors have demonstrated 10^3 cells/ml. Problems include nonspecific binding, degradation of the antibodies over time, reproducibility of the antibodies, and whether the target can be produced in pure culture to provide a monoclonal antibody. There is also a problem with cross-reactivity, that is, closely related organisms frequently cannot be distinguished by immunochemical techniques. In addition, some viruses possess hypervariable coat proteins, and a monoclonal antibody raised against a particular coat protein of a virus may be totally useless for detection of the same virus after it has been propagated for several generations. Nevertheless, some of the most sensitive sensors are based on antibody probes, and a recent variation called immunoPCR that tags the antibody with a short strand of DNA takes advantage of PCR amplification of the antigen-antibody complex to increase sensitivity still further (Joerger et al., 1995). E. I. DuPont Co. and USAMRIID are attempting to apply this technology to simultaneous detection of multiple threat agents.

Ligand-based probes were developed on the principle that every cell has cell-surface proteins that bind other specific molecules. Ligands may be small or large, specific to a particular microbial serotype or common to related groups, and bind with varying degrees of affinity.

Until the recent development of combinatorial chemistry methods, ligand-based probes directed at specific receptors had been dyes that are structural analogs for ligands of microbial receptors and used in classical microbiological screening tests. More recently, scientists at Utah State University (Powers and Ellis, 1998) have capitalized on the fact that pathogenic bacteria, and only pathogenic bacteria (at least the >40 bacterial pathogens they have tested to date), bind hemin to produce a bacterial pathogen detector that, while not identifying the pathogen by genus and species, will detect as few as 100 pathogens in a sample containing ten million or more nonpathogenic bacteria. Researchers at the University of Alabama Birmingham (Turnbough and Kearney, 1998) have screened a library of random 7-amino acid peptides to identify a peptide ligand that binds very strongly and specifically to the spore coat of the nonpathogenic bacterium *Bacillus subtilis*. A similar strategy is to be employed to find a tight-binding ligand for spores of *B. anthracis* and other biological agents. Other ligands include microbial adhesins and oligosaccharides. The Utah State University researchers are now using a variety of combinatorial libraries to find ways to "capture" the toxins produced by *B. anthracis, C. botulinum, S. aureus*, and numerous other pathogenic microbes (Powers and Ellis, 1998). Two potential advantages of this approach are that several toxins may operate by similar mechanisms and therefore may be detectable with the same ligand and that a toxin-based probe will be useful even if inventive weaponeers find a way to deliver a known toxin with a bioengineered organism or a common and ordinarily harmless microbe. Difficulties encountered in developing ligand/receptor probes are interference and competition with natural ligands, as well as the fact that receptor sites are under gene regulation that may alter the expression state in various environmental conditions.

Transducer Technologies

Transducer technologies include: electrochemical, piezoelectric, colorimetric, and optical systems. The transducer system must acquire signals that are unique to the probe system and generate low noise signals that can be further processed without degradation to provide a human observer with an indication of probe system activity.

Electrochemical transducers utilize enzymes to generate an electrochemical signal, either amperometric or potentiometric (amperometric sensors are more sensitive). Commercial examples include sensors for glucose, lactose, and a host of cell products. A Navy-funded R&D effort at Northwestern University is the only example of this approach in our inventory. The principal investigator hopes to immobilize redox-active oligonucleotides on a film in such a way that only sequence-specific hybridization can carry current through the film. Stability is affected by usage and nature of the probe. Response and recovery times are primarily dependent on the rates of diffusion from target to probe reaction sites and from product to electrode. Measurements of 10^3 microbes/ml have been demonstrated in 1–3 minutes.

Piezoelectric transducers rely on the use of certain crystals that produce an electric charge when subjected to pressure. Subjecting those crystals to an electric current causes them to vibrate at a frequency that is dependent upon their dimensions, including their mass. Coating the surface of the crystal with, for example, antibody or nucleic acid probes will alter that frequency, and more importantly, antigen binding or nucleic acid hybridization will cause still more frequency change. There are some problems with reproducing surface coatings, and the sensitivity is typically 10^5–10^6 cells. Specificity is derived from the probe material. Piezoelectricity is the basis for several chemical agent detectors using surface acoustic wave (SAW) technology, and it has been a popular approach to biodetection in the recent past (Guilbault and Schmid, 1991; Guilbault, Hock, and Schmid, 1992). Our inventory of active research shows only two such entries, however: a NASA-funded contract at Southern University to develop a liquid-phase crystal immunosensor, and a Naval Research Laboratory effort to develop an antibody-based force amplified biological sensor (FABS). Both are still in the proof-of-principle stage, the former using *E. coli* for prototype development, and the latter MS2 virus and *B. globigii.*

Light absorption, or *colorimetry*, has also been used for transduction. A binding event causes a color change that can be observed by the naked eye and/or quantified by spectroscopic measurements. For example, colloidial gold bound to agent-specific antibodies produces a red spot when "collected" by antigens in the sample. The "litmus test" being developed by Charych and colleagues at the Lawrence Berkeley National Laboratory is another colorimetric assay. Ligands that bind to specific viruses and toxins are incorporated into a polymerized bilayer assembly that changes color when the agent binds (Charych et al., 1996). This quick and simple test has a sensitivity of 10^8 virus particles and 20 ppm for toxins. Although the sensitivity of colorimetric methods in general is significantly less than that achieved with fluorescence, such methods are useful where the agent is likely to be present in high concentrations.

Optical transduction is employed in the majority of the biodetectors listed in Appendix B. Although a variety of methods based on light scattering and absorbance have been explored in other settings, nearly all the optical examples in our inventory involve fluorescence and other luminescence spectroscopies. Fluorescence approaches involve excitation of the molecules of a material with light, usually in the ultraviolet (UV) portion of the spectrum. The excited component spontaneously reverts to its unexcited state, a process accompanied by emission of light at different wavelengths. These emission wavelengths are dependent upon both the exciting wavelength and the molecules being irradiated, so it is possible to use the resulting emission spectrum to identify the irradiated material. Many biological materials, for example tryptophan, are naturally fluorescent. Due to a number of factors, including the presence of common substances like tryptophan, the luminescence characteristics of many biological and environmental substances overlap—often making identification difficult, if not impossible. However, a variety of methods have been developed to separate individual contributions and the background. Of particular importance are wavelength and phase modulation, as well as time-correlation and line-shape fitting methods. A related indirect approach involves introducing a special fluorophore (a fluorescing chemical with a distinctive emission spectrum) into the sample or the probe molecule prior to irradiation. Ultimately, background and scattering limit the sensitivity and overlapping substances limit specificity. Regardless, optical methods offer the highest sensitivity and selectivity and have been the only methods used for research requiring single-molecule detection.

Two variants of fluorescence being utilized in DoD research on bioagent detection are up-converting phosphor technology (UPT) and the fiber optic evanescent wave guide (FOWG). The former, whose development is being funded through DARPA (Wollenberger et al., 1997; Wright et al., 1997; Cooper, 1998), uses a number of rare earth compounds that, in crystal form, have the unique property of emitting a photon of visible light in response to absorbing 2 or 3 photons of lower-energy infrared light of the proper wavelength. Coating the crystals with antibody provides for a highly identifiable signal, since no naturally occurring substances upconvert. Nine spectrally unique phosphors have been synthesized to date, making it possible to simultaneously probe with as many as 9 antibodies. More phosphors are under development, although it seems likely that the multiplexing limit will probably be closer to 9 than to 100.

The Analyate 2000 fiber-optic evanescent waveguide biodetector developed by the Naval Research Laboratory (Cao et al., 1995; Anderson et al., 1996) also uses antibody probes, some of which are bound to a glass optical fiber immersed in a capillary tube containing an aqueous solution of the sample. Other antibodies, tagged with a fluorescent dye, are added to the sample, where they bind to the target antigen. The antigen-labeled antibody complex then binds to the immobilized antibody. Light from a near infrared diode laser travels through the fiber, which contains it almost completely. The very small amount of light escaping, the evanescent wave, excites the fluorescent tag, whose emission is sent back up the fiber and detected via photodiode.

Hybrid Technologies

There are some detection devices in which there is no clear division of probe and transducer. Methods based on physical properties and separation are good examples: mass spectrometry and gas or liquid chromatography. Mass spectrometry (MS) is a major analytical technique in which materials to be analyzed are converted into gaseous ions or otherwise characteristic fragments. The fragments are then separated on the basis of their mass-to-charge ratio. A display of this separation constitutes the mass spectrogram. MS is an extremely sensitive, selective, and rapid technique. Quantities of chemicals as small as 10^{-18} moles can be detected within milliseconds in highly purified samples, and MS has demonstrated detection of 10^6 cells. In a field environment, or whenever samples are heterogeneous, the constituents must be separated before they can be reliably identified, a task accomplished in a variety of ways, including gas chromatography (GC), high performance liquid chromatography (HPLC), or the use of two mass analyzers (one to perform the separation and a second to produce the mass spectrum of the resulting analytes).

Another separation-based detector system specifically for viruses is being developed by the Army's Edgewood Research, Development and Engineering Center (ERDEC). Based primarily on sedimentation rate with ultrafiltration, proven technologies, the device uses an ultracentrifuge and a series of passes through an ultrafilter to separate viruses from the fine solids onto which they are typically adsorbed and from other nonviral background materials. The final stage of the detection process involves electrospray aerosolation of the filtrate, differential mobility analysis, and a condensation nucleus counter to quantify the viruses present. ERDEC recently licensed a commercial partner (EnViron) to continue development and field testing of the device, which they claim will detect and identify all viruses within an hour, with sensitivity as low as 1000 virus particles even in air or liquid with very high levels of contaminating dust, bacteria, protein, pollen, and fungi. The system accepts both air and liquid samples, including blood, without pretreatment (Wick et al., 1997, 1998).

DETECTION OF BIOLOGICAL AGENTS IN THE ENVIRONMENT

Real-time detection and measurement of biological agents in the environment is daunting because of the number of potential agents to be distinguished, the complex nature of the agents themselves, and the myriad of similar microorganisms that are a constant presence in our environment and the minute quantities of pathogen that can initiate infection. Few, if any, civilian agencies at any level currently have even a rudimentary capability in this area. A number of military units, most notably the Army's Technical Escort Unit, the U.S. Marine Corps Chemical Biological Incident Response Force, and the Army Chemical Corps, presently have some first-generation technology available.

For example, the Biological Integrated Detection System (BIDS) continuously samples ambient air and determines the background distribution of aerosol particles. Aerosol particles with diameters in the 2 to 10 micron range are concentrated and analyzed for biological activity, as indicated by the presence of adenosine 5'-triphosphate (ATP). Flow cytometry then separates and concentrates bacterial cells, and antibody-based tests are conducted for specific agents. At present, the system includes tests for the bacteria responsible for anthrax and plague, botulinum toxin A, and staphylococcal enterotoxin B.

Much less expensive point detectors are available as prototype "One Step Hand-Held Assay" devices. These instruments are currently produced by the Navy Medical Research Institute (NMRI) at Bethesda, Maryland, (similar devices have recently become commercially available through Environmental Technologies Corporation) and are based on antigen capture chromatography. Eight different devices are used to assay liquid samples for the presence of *Y. pestis*, *F. tularensis*, *B. anthracis*, *V. cholerae*, SEB, ricin, botulinum toxins, and *Brucella* species, respectively. A color change provides a positive or negative indication within 15 minutes. The sensitivity of these assays varies from an order of magnitude below a fatal dose (ricin) to more than an order of magnitude above the infectious dose (anthrax). These devices are strictly screening assays, and the analyses are subject to error from the introduction of other contaminants. Therefore, positive results need to be confirmed with standard microbiology assays, conventional immunoassays, or genome detection via polymerase chain reaction (PCR) technology. Both NMRI and USAMRIID at Ft. Detrick, Maryland, have deployable field laboratories that can perform these additional confirmatory assays (and assays for 15 to 20 other potential agents). However, the confirmatory assays do not yield results as quickly. Detectors with higher sensitivity than those presently available will be needed to detect biological aerosols at minimally hazardous concentrations.

Potential Advances

Implicit in the three-stage approach to agent identification by the BIDS is the realization that in some circumstances one need only know that there are more particles in air than normal to take some important action, such as put on a respirator. In other circumstances, one might need more information about the nature of the particles (are they biological, and if so, are they living?) to take action. In still other circumstances (forensics or treaty verification), one needs to be able to identify a specific bacterium or virus.

The perceived need for faster, surer results for timely detection of hazardous biologicals in the environment has spawned a large and growing number of research programs. Biological detection is the largest single category in the committee's inventory of relevant technologies (Appendix B). Space does not allow discussion of each, but most of the devices are variations on a small number of approaches that were described in the previous section on patient diagnostics. As in the case of chemical detectors, the underlying approach largely determines the sensitivity, selectivity, versatility, and reliability. Application to detection of biological agents in the environment differs from patient diagnostics primarily in the increased need for portability, ease of use by nonscientists, speed, and methods for collecting and preparing the sample. The following pages first describe the main approaches to sampling the environment for biological agents. We then consider some current research and development on new and better devices for detecting, identifying, and quantifying biowarfare agents and how they might meet needs of civilian medical personnel in domestic terrorism scenarios. We conclude with recommendations for prioritizing R&D in this area.

Sampling

Sampling has to do with how the material that is to be tested is brought to the detector, whether it comes from air, liquid, solid objects, surfaces, or from human tissue. There are several issues that make sampling for biological agents challenging. The first issue is that the sampling is normally targeted at living organisms; therefore, the technology must not "harm" the sample. Secondly, because most detector devices require a liquid sample, collection of airborne microbes must be extracted from an aerosol or particulate for and concentrated in a liquid. Third, the target microbe is generally only one component of a complex matrix of biological elements and chemical compounds that may affect the detection process, so the sample must often be purified to some extent. Last, the sample must be highly concentrated for a rapid analysis. Four general types of sampling devices designed to accomplish one or more of these objectives are: (1) viable particle-size impactors, (2) virtual impactors, (3) cyclone samplers, and (4) bubblers/impingers. Each of these technologies is described below.

Viable Particle-Size Impactors. The viable particle-size impactors usually have multiple stages. Each stage contains a number of precision-drilled orifices that are appropriate for the size of the particles to be collected in that stage, and orifice sizes decrease with each succeeding impactor state. Particles in the air enter the instrument and are directed towards the collection surface by the jet orifices. Any particle not collected by that stage follows the stream of air around the edge of the collection surface to the next stage. The collection plate is typically a petri dish with agar or other suitable growth medium (Boiarski et al., 1995).

Virtual Impactors. A virtual impactor is similar to a viable particle-size impactor, but uses a collection probe instead of a flat plate as its impaction surface. Air flows through the collection probe and the collected particles are transported to other portions of the collector for additional concentration. By controlling the flow in the impactor, it is possible to adjust the cutoff size to the particles collected. By passing the collection probe airflow into successive virtual impactors, the particles can be concentrated to many times the original air concentration before collection. The final stage can then impact the particle stream into a liquid, resulting in a highly concentrated liquid sample (Boiarski et al., 1995).

Cyclone Samplers. A cyclone is an inertial device that is commonly used in industrial applications for removing particles from large air flows. A particle-laden air stream enters the cyclone body and forms an outer spiral moving downward towards the bottom of the cyclone. Larger particles are collected on the outer wall due to centrifugal force. Smaller particles follow the airstream that forms the inner spiral and leave the cyclone through the exit tube. Application of a water spray to the outer walls of a cyclone facilitates particle collection and preservation. (Boiarski et al., 1995).

Bubblers/Impingers. Most bubblers or impingers operate by drawing aerosols through a current inlet tube and jet. Usually the jet is submerged into the liquid contained in the sampler. As the air passes through the liquid, the aerosol particles are captured by the liquid surface at the base of the jet. In order to collect the smallest particles possible, the jet is typically made with a small critical orifice causing the flow to become sonic. Other designs have a fitted jet so that tiny air bubbles are formed in the liquid as air leaves the jet. (Boiarski et al., 1995).

Two very important sampling issues must be addressed, regardless of the technology employed. First, the environment in which the target microbe exists can significantly affect the physiology of the microbe and with that the efficacy of the detection procedure. *Bacillus anthracis*, the causative agent in anthrax, provides a simple example: in the environment it exists as a hard, oval, inactive spore highly resistant to sunlight, heat, and disinfectants, but in tissue, including blood, it germinates into a rod-shaped vegetative bacillus actively proliferating and producing its characteristic toxins. Detection strategies appropriate for one form of the organism may be entirely ineffective in the presence of the other form. Less dramatic but equally important, components of the matrix in which a microbe exists contribute significantly to the microbe's growth state and gene expression in a way that is just beginning to be explored for most organisms. Detection strategies focused on a specific structure or gene product can thus vary wildly, if sampling conditions are not clearly specified.

The second overarching sampling issue is especially important in attempts to detect microbes in very low concentrations: the process is a statistical problem, and due consideration must be given to the variables that affect any statistical conclusion, namely the size, number, randomness, and independence of the samples.

Stand-off Detection

If there is advanced warning of an "event," then diagnostic capability requirements also include not only "point" detection (in which the detector directly samples the contaminated environment), but also real-time "stand-off" detection (detection is accomplished from a distance) Because most of the agents under consideration in this document are considered attractive as weapons in part because they can be delivered as aerosols, DoD is developing "stand-off" monitors aimed at detecting particles of a biological nature in distant clouds. The simplest of these optical devices merely looks for unexplainable increases in the thermal emissions from a given direction, but the more sophisticated uses ultraviolet laser-induced fluorescence to identify the presence of tryptophan. Current prototypes are a large improvement over earlier stand-off systems, but they cannot yet consistently identify specific organisms because of the similarity of their emission spectra. Advanced signal processing techniques may improve identification.

Sensitivity to infectious dose level is probably not important for early warning, since an aerosol cloud intended to kill or incapacitate even one individual will certainly involve concentrations far in excess of the infectious dose (later decisions about clean-up and reoccupation of contaminated areas may need that level of sensitivity, but speed will be less of an issue, and respiratory protection will allow use of more sensitive point detectors). Specificity also may not be critical in the use of stand-off detectors. For example, we may just need to be alerted to the presence of live biologicals. This is also true for the control of contaminated environments, determination of decon efficacy, and dynamic threat assessment (real-time assessment of a threat, including remediation).

Stand-off detection offers safe, real-time determination of microbial contamination. Significant advances have been made with the use of lasers for the detection of aerosolized agents by light-scattering characteristics, infrared and Raman spectroscopy, and fluorescence, but these same methods can also be used to determine total microbial contamination on objects (Powers and Ellis, 1998) and in situations where effective sampling is impossible for reasons other than distance. The efficacy of these devices is somewhat limited by the range at which the determination is desired (typically several kilometers for military systems). Longer distances, of course, are more difficult, and the necessity for prior intelligence, subsequent deployment, and then line-of-sight use of the technology would seem to limit its utility in urban bioterrorism scenarios. Applications of true stand-off detection would seem to be limited to monitoring predetermined, high-risk sites or large public gathering places, such as stadiums, for aerosol clouds.

An alternative approach for long distance detection is the small model airplane-like unmanned aerial vehicles (UAV) being developed by the Naval Research Laboratory and Research International (Foch, 1998). These vehicles, ranging in size from a few inches to a foot in size, may eventually carry on-board sensors and down-link data to ground-based control. There is a weight and size limitation on the sensors that can be carried on-board, but prototype vehicles have been successfully demonstrated in cities and inside buildings as well as in outdoor terrain. Furthermore, they are reusable and easily transported. In the event of biological agents being released in a building, such vehicles could locate "hot zones" and monitor decon efficacy, reducing human exposure and risk.

Point Detection

Point detection refers to testing a sample that has been taken directly from the environment suspected of harboring the target agent. Needs in this regard include not only investigation of suspected sources of contamination but also monitoring the air/water systems in buildings for general pathogen contamination or contamination by specific biological agents. A number of the embryonic microbe detectors described above in the section on patient diagnostics are being examined for utility in environmental detection as well.

In point detection as in stand-off detection, many situations will demand neither exceptional sensitivity nor exceptional specificity. Assessment of the total microbial content may be sufficient to determine contamination and alert personnel to danger. For example, if there is already a suspicion that a terrorist attack is likely, then a sharp and unexplainable rise in total microbial count probably should be sufficient to trigger protective action, regardless of whether the specific pathogen can be identified. Total microbial count might also be sufficient for the assessment of decontamination efficacy. In other situations, perhaps a detection situation when information is not available on "background" microbial levels, knowledge of total pathologic organisms present may be sufficient to guide short-term actions by rescue and medical personnel, even if the specific pathogen is not identified. More precise identification would be important for forensic uses of course, and for optimal treatment of many agents (e.g., broad-spectrum antibiotics might be prescribed as soon as the agent is identified as bacterial, even if the species is unknown, but this practice contributes to the development of resistant strains; the few anti-viral drugs available have thus far proven to be virus-specific).

Most of the current R&D on detection of biological weapons employs nucleic acid- or antibody-based probes combined with optical, most often fluorescence, transduction, or it involves adapting separation-based technology like mass spectrometry.

Regardless of the transducer technology employed with nucleic acid probes, "amplification" is generally required to detect the very low number of microbes that suffice to infect humans. A distinguishing feature of nucleic acids is the possibility of rapidly multiplying ("amplifying") distinctive nucleotide sequences in samples too small to be analyzed by other methods. This is accomplished by enzymatic [polymerase chain reaction (PCR), ligase chain reaction (LCR), Q-beta replicase] or nonenzymatic methods, such as Chiron Corporation's HIV RNA assay using a covalently branched DNA structure. All of these methods separate a piece of the normally double stranded DNA into constituent single strands, each of which, given the necessary amino acids, assembles a complementary strand, the net result of this "cycle" being a doubling of the number of target DNA strings.

The sensitivity of detection of nucleic acids can thus be greatly improved by nucleic acid amplification. The polymerase chain reaction (PCR) takes time, and a major aim of current research is to shorten the time to approach real-time amplification. Idaho Technology's LightCycler, one of the fastest presently on the market (Wittwer et al., 1997), can carry out 30 cycles in 6 minutes by using tiny glass capillary tubes for the sample and high-velocity hot and cold air. RNA can be converted to cDNA by reverse transcriptase (RT) and thus amplified by PCR. The time required for conversion to cDNA is also a subject of active research. Several new amplification methods do not require heat cycling. These include Transcription-based Amplification System (TAS) (Kwoh et al., 1989), Self-Sustained Sequence Replication (Guatelli et al., 1996), and Strand Displacement Amplification (SDA)(Walker et al., 1992).

In general, however, degradation of the nucleic acid probes and interference from related sequences or products from the microbial environment significantly limit the current application of this technology beyond well-equipped and experienced laboratories. A single microbial cell can be detected in the laboratory from highly purified DNA by these methods, but environmental samples have regularly failed to achieve this, usually having a detection limit of 10^5 microbes. PCR detected 100 percent of spiked samples in one study (Candrian , 1995), but only 15 percent of naturally infected samples. Considerable effort is being made at Lawrence Livermore National Laboratory (LLNL) to solve these problems and combine nucleic acid-based assays with antibody-based tests in an automated field-deployable system (Mariella, 1998; Belgrader, 1998). Miniaturized PCR units with significantly reduced cycling times have also been developed by a partnership of USAMRIID, LLNL and the California biotech company Cepheid, Inc. (Ibrahim et al., 1998; Belgrader et al., 1998; Northrup et al., 1998). The long-term goal of this work is a hand-held instrument featuring disposable cartridges containing all necessary reagents, reaction chambers, waste chambers, and microfluidics to extract, concentrate, amplify, and analyze nucleic acids. Concurrent efforts at sequencing the genes of possible biological warfare agents and identifying organism-unique probes are under way at Army (USAMRIID) and Navy (NMRI) laboratories (Farchaus et al., 1998; Higgens et al., 1998), LANL (Keim et al., 1997), LLNL (Andersen et al. 1996), the University of Texas-Houston (Hoffmaster and Koehler, 1997), and Duke University (Harrell et al., 1995), so piggybacking onto a commercial market that Cepheid estimates at over $1 billion seems feasible.

The one-step hand-held tickets described above that are produced at NMRI and more recently by Environmental Technologies Corporation are an example of *immunoassay* technology combined with chromatographic transduction. The sensitivity of these simple devices is much lower than that achieved in clinical laboratories, but they are inexpensive and easy to use. For those reasons, they are probably the most logical choice for Hazmat teams and other emergency responders seeking to test the contents of a suspicious package for the presence of the dozen of so agents on the military threat list. The Analyte 2000 is another well developed (but not yet commercially available) immunosensor. The Naval Research Laboratory developed this device, which combines antibody probes with a fiber-optic waveguide transduction system. Other work is focusing on miniaturizing and automating the testing process, incorporating the requisite antibodies with optimal sampling and transducer technology, and producing antibodies against specific biological agents and strains. Scientists at the University of Texas, Austin (Daugherty et al., 1998; Georgiou and Iverson, 1998) are taking the last of these areas one step further, reducing the size of anthrax antibodies to that fragment of the light chain actually binding the antigen, identifying the relevant amino acids at the binding site, and making systematic substitutions to achieve higher affinity and selectivity.

The previously described device under development at Utah State University for detection of total pathogenic microbes (including spores) is an example of a *ligand-based* probe with fluorescence-based transduction. A hand-held unit simply using fluorescence to determine total viable microbes requires no physical contact with the samples and no specialized expertise to use, but it can provide detection in seconds with a sensitivity of ~100 cells. In this respect, it is useful for determining contamination on objects and from environments where it is difficult to obtain samples.

Three substantial R&D efforts are currently under way that focus on mass spectrometry (MS) for identifying biological agents. DoD is close to fielding a truck-portable Chemical Biological Mass Spectrometer (CBMS) and already has research under way at Oak Ridge National Laboratory (ORNL) for a second-generation unit that is lighter, faster, and more sensitive (Wayne Griest, personal communication to FJ Manning, 1/23/98). Although very expensive compared to most portable chemical or biological detectors and dependent on a rapid and efficient separation system, the name underlines an important advantage of this approach—the potential for a single instrument that will detect both chemical and biological agents, industrial and naturally occurring as well as military. Unlike many of the current test systems and detectors, such an MS-based detector could be used in a whole gamut of Hazmat situations rather than as confirmation of a hypothesis about a possible agent. The instrument's versatility would be limited only by the size of the existing library of mass spectra.

DARPA is sponsoring a collaboration of Johns Hopkins University, the University of Maryland, and USAMRIID to develop a portable, fully automatic MS system and a library of bioagent "signatures" (Cotter, 1998; Fenselau, 1997, 1998; Bryden et al., 1998). The Department of Energy's Chemical Biological Nonproliferation Program is sponsoring a similar developmental effort at ORNL, where researchers are attempting to leverage hardware and software engineering currently under way in connection with the second generation CBMS to produce a man-portable, real-time system capable of identifying airborne bacteria or volatile organics as well as characteristic proteins of biowarfare viruses, toxins, and bacteria (McLuckey, 1998; McLuckey at al., 1998; Stephenson et al., 1998).

Although MS has the potential to identify infective agents and recent advances have significantly reduced the size of the device, libraries of unique signatures of agents have not been determined. In addition, it is not clear that these signatures can be distinguished in a natural environment containing signatures of large amounts of other microbes, especially at concentrations near infectious-dose levels.

Other detectors being developed at Sandia National Laboratory are based on miniaturizing standard laboratory separation techniques, such as capillary zone electrophoreses, size exclusion chromatography, and reverse phase and affinity electrochromatography coupled with fluorescence (Vitko and Kottenstette, 1998; Dulay et al., 1995; Ramsey et al.,

1995). The challenge with these technologies is to achieve high sensitivity in the presence of large amounts of interfering substances. Interfering substances may have the same physical parameter that is being used for selectivity, such as, charge, size, mass, which can cause wrong results, even though the results are highly reproducible. For that reason, the investigators propose to use as many as four of these techniques in parallel. Only when a sample is positive on all methods would the result be considered unequivocal.

R&D NEEDS

The type of detection technology that is needed depends upon the scenario, and, as is the case with chemical agent detectors, it is likely that no one detector will meet all civilian needs. As with R&D needs in other parts of this report, detector technology needs were evaluated with three scenarios in mind: (1) general monitoring in the high-risk environment, (2) an "event" (most likely a suspicious package in the case of biological agents, but possibly an explosion of some sort), and (3) a "covert" release (patient diagnostics). The first two scenarios call for some ability to detect biological agents in the environment (air, water, food, etc.), while the third calls for methods that will detect and identify pathogens in fluids or tissues from patients who exhibit signs or symptoms, or who are known to have been exposed to a pathogen.

The committee does not see routine monitoring in the manner of smoke detectors (i.e., without some independent reason to suspect an attack) as either feasible in the foreseeable future or worthy of a high-priority effort to develop that capacity, but there may be times and places where pre-incident intelligence may suggest temporary deployment of existing military monitoring systems.

Given the delayed effects of the biological agents, it is also difficult to envision many situations that would demand highly sensitive biological detection by first responders. The ability to determine total viable microbes present, total pathogenic microbes, and specific viable pathogens will likely cover the needs presented by both overt and covert "events" as well as provide monitoring and early warning. The ERDEC/EnViron virus detection system might prove to be a useful complement to a ligand probe system for detecting total pathogenic bacteria and handheld immunoassay tickets in a multistage approach beginning with the very general and progressing to the highly specific as required. The alternative might be a miniaturized mass spectrometer of the sort being developed at Hopkins/Maryland or Oak Ridge to be a generic chemical and biological identifier. Although prototypes are decreasing in size and weight, the real challenge lies in the development of a library of unique signatures for biological agents in the presence of large quantities of other microbial contamination and interferents, in addition to the achievement of infectious-dose-level sensitivity.

In the area of patient diagnostics, there is a clear need for methods capable of detecting infective dose levels (e.g., 10–100 cells or virions) of most biowarfare agents at a speed that allows for effective therapeutic strategies to be administered (e.g., antibiotics, vaccination, supportive therapy). Furthermore, these new methods must also be able to detect "friendly" microbes that have acquired virulence factors by natural or genetic engineering methods and those that have been microencapsulated to disguise their identity (such as the detection of virulence factors or toxin production). Ideally, this technology will be incorporated into a diagnostic system capable of identifying many more common pathogens, assuring frequent use of the system and eliminating the need for clinicians to make a specific request for a very seldom-used assay.

The committee therefore has identified the following research and development needs:

6-1 In the area of patient diagnostics, the Public Health Service should encourage federal research agencies to leverage burgeoning commercial development of faster, cheaper, easier assays of common pathogens rather than independently developing diagnostic technology for the less common pathogens thought to be good candidates for bioterrorism.

6-2 In the area of environmental detection, the Public Health Service should closely monitor military biodetection R&D efforts for inexpensive or multipurpose biodetectors that might be appropriate for purchase or loan by civilian agencies rather than developing threat agent-specific assays from the ground up.

6-3 Both of these leveraging efforts will require the federal government to conduct or support:

• Basic research to identify characteristics which might be used to develop more effective probes and/or enhance probe performance for known biowarfare agents and especially genetically altered microbes. Understanding of microbial metabolism, sporulation, toxin production and excretion, regulation of virulence factors, and bacteriophage interaction are crucial in this respect. New approaches for preventive and therapeutic strategies are also likely from this basic understanding.

• Scenario-specific testing of detection performance and comparisons under standard conditions for characterization of the sensitivity, specificity, reliability, response constraints, and usability (ease of use, cost, robustness, useful life, response time, and human effort and experience required).

7

Patient Decontamination and Mass Triage

Decontamination is defined as the process of removing or neutralizing a hazard from the environment, property, or life form. The principal objectives of this process are to prevent further harm and optimize the chance for full clinical recovery or restoration of the object exposed to the dangerous hazard. The triage process is the initial step taken to meet the primary objectives of a disaster response, which involves sorting the injured by priority and determining the best utilization of available resources (e.g., personnel, equipment, medications, ambulances, and hospital beds). This chapter includes a review of decontamination and mass triage with an emphasis on the research and development needs in these areas of disaster response.

DECONTAMINATION

Fire departments and hazardous material teams have traditionally described the decontamination processes with two terms—"technical decon" and "medical" or "patient decon." "Technical decon" is the process used to clean vehicles and personal protective equipment (PPE) and "medical" or "patient decon" is the process of cleaning injured or exposed individuals.

Technical Decontamination

Technical decon is most commonly performed using a sequential nine-step process originally developed by Noll and Hildebrand (1994). The steps are listed below.

In the Exclusion Zone (Hot Zone—dangerous concentrations of the agent are likely)
1. Contaminated tools and equipment drop onto a plastic sheet
2. Contaminated trash drop

In the Contamination Reduction Zone (Warm Zone)
3. Primary garment wash/rinse (boots, outer gloves, suit, SCBA, and mask)
4. Primary garment removal
5. Secondary garment wash/rinse (decontaminate inner protective garment and inner gloves)
6. Face piece removal/drop (can be combined with stations 7 and 8)
7. Boot drop
8. Inner glove removal

In the Support Zone (Clean Zone)
9. Shower and clothing change

This process is well known and extensively utilized by the public safety community. Cleaning is done using water in conjunction with one of four cleaning solutions, (solutions known as A, B, C, D), depending on the type of contaminant. Solution "A" contains 5 percent sodium bicarbonate and 5 percent trisodium phosphate and is used for inorganic acids, acidic caustic wastes, solvents and organic compounds, plastic wastes, polychlorinated biphenyls (PCBs), and biologic contamination. Solution "B" is a concentrated solution of sodium hypochlorite. A 10 percent solution is used for radioactive materials, pesticides, chlorinated phenols, dioxin, PCB, cyanide, ammonia, inorganic wastes, organic wastes, and biologic contamination. Solution "C" is a rinse solution of 5 percent trisodium phosphate. It is used for solvents and organic compounds, PCB and polybrominated biphenyls (PBB), and oily wastes not suspected to be contaminated with pesticides. Solution "D" is dilute hydrochloric acid. It is used for inorganic bases, alkalis, and alkali caustic wastes.

Once the decon process is completed, the equipment is most often returned to service, unless the item(s) cannot be completely decontaminated (as determined by using available detection devices). However, current research does not provide an answer to the question, "how clean is clean?" Some communities will depend on disposable equipment as an alternative to trying to assure that each item has been thoroughly decontaminated. Other communities may not be able to afford the replacement cost and depend on using available technology or best guess to determine when these items are "clean." It will be important for emergency responders to know when technical decontamination has been achieved, if the equipment is to be reused. It is vital when personal protective clothing or equipment is involved.

Patient Decontamination

Patient decontamination, which Hazmat teams have to undertake much less often than technical decon, is to be performed when the contaminant poses a further risk to the patient or a secondary risk to response personnel. Fire and EMS publications frequently describe how patient decontamination can be done, but few of the recommendations are based on empirical research. Because little scientific documentation exists for when and how patient decontamination should be performed expeditiously and cost effectively, prehospital and hospital providers are left to doing what they think is right, rather than doing what has been proven to work best. Generally, the process involves three stages; gross, secondary, and definitive decontamination.

Gross Decon
1. Evacuate the patient(s) from the high-risk area.
2. Remove the patient's clothing.
3. Perform a one-minute quick head-to-toe rinse with water.

Secondary Decon
1. Perform a quick full-body rinse with water.
2. Wash rapidly with cleaning solution from head to toe.
3. Rinse with water from head to toe.

Definitive Decon
1. Perform thorough head-to-toe wash until "clean".
2. Rinse with water thoroughly.
3. Towel off and put on clean clothes.

As noted above, among the first steps in the decontamination process is the removal and disposal of clothing. Cox (1994) estimates that 70 to 80 percent of contaminant will be removed with the patient's clothes. Little scientific data exist to support this assertion, however. The ideal skin decontaminant would remove and neutralize a wide range of hazardous chemicals, be cheap, readily available, rapid acting, and safe. For most civilian applications, water has been the choice; the technical decontaminant solutions cannot be safely used to clean the skin or mucous membranes. The armed forces have assessed a wide variety of skin decontaminants, including flour, Fuller's earth, and absorbent ion-exchange resin for environments where water is not available. A fresh solution of 0.5 percent sodium hypochlorite appears to be the state-of-the-art liquid decontaminating agent for personnel contaminated with chemical or biological agents (Chemical Casualty Care Office, 1995). The half-life of sarin in undiluted household bleach, which is 5.0 percent sodium hypochlorite and generally too harsh for use on skin, is on the order of 3 seconds (Kingery and Allen, 1995).

Civilian Hazmat teams generally have basic decontamination plans in place, though proficiency may vary widely. Very few, if any, teams are manned, equipped, or trained for mass decontamination, however. Again, water is the principal decontamination solution, with soap recommended for oily or otherwise adherent chemicals. Some teams suggest that initial mass decontamination be accomplished by fire hose (operated at reduced pressure), which has the advantage of being possible even before the Hazmat team arrives on scene (the MMST equipment list includes hoses specifically for this purpose). Shower systems with provisions for capturing contaminated runoff are commercially available and may provide some measure of privacy in incidents involving only a handful of victims (they generally accommodate only one person at a time). However, the availability of trained personnel in appropriate personal protective clothing is likely to be a limiting factor, even when larger shower units or multiple smaller ones are available. The CBIRF and MMST have much larger shower units, capable of decontaminating dozens to hundreds of victims with sodium hypochlorite solution, and are staffed at much higher levels than local Hazmat teams. However, neither will be immediately available unless predeployed (as was done, for example, at the Atlanta Olympics and State of the Union Address). Harsh weather, intrusive media, and the willingness of ambulatory patients to disrobe in less than private surroundings will also affect the conduct of field decontamination. Where there are very large numbers in need of decontamination, crowd control measures will be necessary to keep panicky or merely impatient victims at the scene long enough to complete decontamination.

The degree to which a patient is decontaminated in the prehospital setting depends on the decon plan, available resources, the weather, and patient volume. At minimum, every patient presenting a risk of secondary contamination risk should receive gross decon before departing for the hospital. These patients should be transported to a hospital (by properly protected EMTs and paramedics). The receiving hospital should be equipped and staffed to perform secondary and definitive decon, if not already done in the field.

Patients requiring additional medical attention, such as attention to the ABCs (airway, breathing, and circulation), antidotes, or other emergency treatment, may receive that care during or after the decontamination process depending on the severity of the agents' effects and the ability of the decon team and available medical personnel to render that care. Nonambulatory patients pose much more of a decontamination and treatment burden than ambulatory patients, because most portable decontamination chambers require a person to stand. Decontamination and treatment planning must also address how to deal with the pediatric patient and the elderly.

Although hospitals are required by the Joint Commission on Accreditation of Healthcare Organizations (JCAHO) to be prepared to respond to disasters, including hazardous material accidents, few have undertaken realistic planning and preparation. Some hospitals have decontamination facilities; however, very few have outdoor facilities or an easy way of expanding their decontamination operations in a mass-casualty event (Cox, 1994; Levitin and Siegelson, 1996). Often their initial response to an incident will be to contact the local fire department or Hazmat team for assistance. This will not be a viable solution if the incident is large or nearby. Unannounced ambulance or walk-in patients who are contaminated may create havoc and harm before "outside" help arrives to address the situation or internal resources can be organized to respond. If assistance from the local public safety agency is not available, the hospital is left to fend for itself and, if unprepared, the response is likely to place the patient, staff, and facility at great risk. There is little financial incentive for a hospital to be prepared for a "once in a lifetime" event, and proper equipment and training may be perceived as too expensive under the circumstances. Generally, hospitals that are prepared are usually capable of handling only a few patients an hour. What happens when a large number of patients begin to arrive? Currently, the medical literature does not contain sufficient research findings to assist hospitals with cost-effective Hazmat or terrorist response planning. The Agency for Toxic Substances and Disease Registry recently released a series of guidelines to help local emergency departments, communities, and other policymakers develop their own response plan or hazardous materials incidents (U.S. DHHS, 1994a), and the Centers for Disease Control and Prevention's Planning Guidance for the Chemical Stockpile Emergency Preparedness Program (CSEPP) provided recommendations for civilian communities near chemical weapons depots (U.S. DHHS, 1995b). Although helpful, the outlines are very generic, do not address how to actually perform mass decon, and do not contain information on many of the agents which are likely to be seen in a terrorist incident. Since planning is left to the local jurisdictions, the success of any national initiative is dependent upon cooperation at the local level.

Aside from the issues related to effective decontamination procedures, training of emergency department personnel must also be considered. There are few courses emergency department personnel may attend to improve their level of preparation for decontamination of large numbers of people.

Potential Advances

Much has been learned about patient decontamination and mass triage in recent years from the process to the equipment. The following section highlights some of these advances and identifies needs for additional research and development.

The Decontamination Process

Further research is needed to determine when decontamination is really warranted and the most effective way to establish and correctly conduct the decon process both in the field setting as well as in the hospital. Both U.S. Army (Chemical Casualty Office, 1995) and FEMA (Federal Emergency Management Agency) guidance suggest that decontamination is unnecessary when dealing with agents in nonpersistent (vapor) form. Under these circumstances, removal of the patient from the source of the vapor is all that should be necessary, and decontamination would needlessly delay evacuation and treatment. In practice, a number of extra-scientific reasons can be adduced for making decon routine: the agent(s) cannot always be identified immediately, medical personnel may be endangered by very small amounts of agent present on each of a long series of patients, and to protect the psychological well-being of both victims and emergency workers.

Recent reports on the Tokyo subway incident of 1995, which involved the nonpersistent nerve agent sarin, provide some support for this position (Okumura et al., 1998a,b). No field decontamination was performed onsite, and emergency medical technicians (EMT) transported 688 victims to hospitals by ambulance. Ten percent of 1,364 EMT showed symptoms and had to receive treatment at the hospital themselves. Once the hospitals (at least St. Luke's) learned that nerve agent was suspected, the most seriously ill patients were directed to a shower upon arrival. Their clothes were placed in plastic bags and sealed up. Despite these precautions, and the use of surgical masks and gloves, 110 hospital staff (23 percent) complained of acute poisoning symptoms on a follow-up questionnaire.

To perform patient decontamination safely and correctly requires a response plan, proper equipment, and trained personnel. Military procedures, and adaptations thereof for use in the CSEPP provide generic guidance for some highly specific situations, but to date there is no detailed national guideline on how to set up and conduct a massive decontamination process in the civilian setting. Ideally, this guideline would address areas such as site management and crowd control, cleaning ambulatory and nonambulatory victims, handling the special needs of pediatric and geriatric populations, and a standardized patient assessment and triage process to be initiated by personnel wearing PPE to determine viability and need for decontamination.

Besides the need for a step-by-step process for performing decon in the field setting and in the emergency department, there is no good way to determine when a patient is "clean." Few chemical or biological agents can be readily seen on the skin or quickly assayed to determine whether any residual product remains after washing. Existing technology is either not available, too expensive, or does not provide the needed versatility to be used in the civilian environment. In the absence of knowing "when clean is clean enough," prehospital and hospital personnel are left to process certification (we followed the SOP, so the person or item must be clean) or using their best clinical judgment as to when the decon process can be terminated—an inefficient, and potentially unsafe, practice in many instances. Affordable, accurate, and durable detection devices that are able to reliably establish that no further clinical risk remains to the patient need to be developed so that emergency personnel will know when a patient is "clean." Of course, once the guidelines and technology are in place, issues of funding for EMS and hospital personnel training will need to be addressed.

Cleaning Agents

The ideal cleaning agent is inexpensive and nontoxic, is rapidly applied and effectively removes the entire contaminant from personnel, equipment, and vehicles. At this time, more is known about technical decon than patient decon, which, as mentioned above, generally involves the use of either soap and water or sodium hypochlorite (0.5 percent). Little research exists to show which soap is best and how long a body surface area must be scrubbed before it is properly cleaned. However, a recent review of the literature by Hurst (1998) suggests that under certain conditions bleach, even at the 0.5 percent level, may actually increase the toxicity of some nerve agents. The M258A1 and M291 are individual skin decontamination kits used by the military and are not routinely available or familiar to the civilian population. Their applicability for use in the civilian setting or in mass decon efforts has not been studied. Current military research on the use of foams, gels, catalytic solvents, and Fenton reagents may have some application for performing technical decontamination in the civilian setting, but more research is needed to determine which agent(s), if any, are suitable for use on civilian patients of all ages and what advantages they have over water or hypochlorite.

Patient Showering Equipment

The equipment currently used by many EMS and hospital personnel during decon is very rudimentary and often "home made." Commercially available equipment is often expensive and designed for technical decon rather than patient decon. For example, containment basins often do not have sufficient size or depth to accommodate patients who are supine on backboards, shower systems correctly wash only standing patients, and patients often stand or lie in the product just washed off them. Patient modesty and protection from the environment are two other problems seen in performing prehospital decon. While some hospitals advertise they have a decon room, often it is too small, or ill equipped to meet its intended purpose. Obtaining large supplies of tepid water can be a challenge for prehospital and hospital decon systems. High-pressure systems require less volume, which helps control runoff as well, but low-pressure, high-volume spray nozzles should theoretically be used to avoid vasodilatation of superficial vessels during rinsing that could enhance agent absorption. However, the necessity of their use has never been scientifically proven. Research on the application of military decon strategy and equipment in the civilian setting has also never been reported. Although the commercial market can certainly produce needed decontamination hardware, development of more standardized methods for conducting patient decon will spur improvements in the suitability and cost of the equipment.

Performing Mass Decon

While the exact number of hazardous material accidents occurring each year may not be known, available data does suggest that for most incidents there are few, if any, injuries (Sullivan and Krieger, 1992). However, terrorists' use of a chemical or biological weapon is likely to lead to scores of injuries and fatalities. The rapid implementation of effective triage and initiation of decon will be vital to optimizing victim survivability and responder safety. But how these two processes should be conducted is neither well known nor extensively studied in the civilian setting. Most hazardous material teams and hospitals have limited experience, usually with five or fewer patients at a time. How they can handle 50, 500, or 5,000 patients in a rapid, efficient, and safe fashion is a critical question being asked across the country. The utilization of an MMST to assist local responders may be part of the answer, but emergency planners and incident commanders must keep in mind it will be 90 minutes or longer before this team (which, in Washington, D.C., for example, consists of 43 members) and its equipment arrives. Federal assistance from DoD will likely take even longer to arrive. Interim solutions will have to be found. Some public safety agencies are using specially designed tractor-trailers to decon multiple patients simultaneously (e.g., New York City). These units can provide protection from the environment as well as privacy from onlookers in addition to deconning multiple patients at a time. However, these trailers are expensive and cannot always be placed in desirable locations within the warm zone. Easily inflatable tents are used as shelters. They provide some of the benefits of trailers and are less expensive, but generally take some time to assemble and cannot handle large numbers of patients at a time. Local communities will need to have a primary decon plan that the first personnel on the scene can rapidly implement and a secondary plan to employ when additional personnel and equipment become available.

Critical to managing the decontamination of large numbers of patients is gaining control of the crowd. Repeatedly giving definitive instructions on what to do over loud speakers is important, along with having an adequate number of properly protected personnel directing the victims through the decon process. Providing verbal instructions may be all that is needed to care for the ambulatory populations, but nonambulatory victims will require more assistance and equipment (e.g., backboards). There is virtually no research being conducted on how to effectively organize and manage such a mass decontamination effort. The military model primarily addresses how to handle young healthy soldiers already wearing protective clothing and respiratory protection, and is not directly applicable to a heterogeneous, unprotected, and undisciplined population. The similar mass decon process envisioned by the MMST has not been utilized except in drills.

Patient resistance to removing their clothing because of modesty or bad weather is a potential problem, but there is no research that validates this issue or its impact. Some suggestions have been made to simply leave the patients' clothing on and spray the crowd with water from hoses located on top of fire apparatus. The effectiveness of this approach, which might actually increase agent-skin contact, has not been studied either.

Organizing a large decon corridor to handle inordinate numbers of patients is another vital concern. Research is needed to determine the optimal responder/patient ratio, how large an area is needed to decon 50,

500, and 5,000 people, what level of medical training is required for the personnel performing decon, and how much medical care should be given in the warm zone as opposed to the cold zone or at the hospital. Delaying or improperly conducting decontamination increases the danger to the patient as well as the health care provider.

No less important is the hospital's ability to process large numbers of victims in a timely fashion. Hospitals need to know how their decon systems should be organized and equipped, whether decon is best done inside or outside of the facility, what PPE emergency department personnel should wear, how the system should accommodate both walk-in and ambulance-delivered patients, and the patient volume that should be manageable in an emergency department that has 10,000, 25,000, or 60,000 visits a year. Another issue is how the cost for being prepared could be recovered by the hospital. Unlike other modernization efforts, a decontamination unit is not going to pay for itself with new patients and fees for the hospital.

Decontamination of Biologic Agents

Biological warfare agents on the skin and clothing of patients pose only minimal risk to medical personnel from aerosolization ("off-gassing") if standard precautions (gown, gloves, eye protection, and careful handling of needles and other "sharps") are observed. Dermal exposure to a suspected agent should nevertheless be treated immediately with soap and water, followed, after a thorough rinse, with a 0.5 percent hypochlorite solution, which will neutralize any remaining microorganisms within 5 to 10 minutes. As noted in the previous section, hypochlorite is contraindicated for decontamination of eyes or in cases of wounds involving brain, spinal cord, or the abdominal or thoracic cavities. Equipment used in caring for potentially contaminated or infected patients should receive special attention in view of the likelihood of its subsequent use with other patients. Normal sterilization with dry heat or autoclaving is ideal, but 30 minutes soaking in a 5.0 percent hypochlorite solution (undiluted household bleach) will serve as a field expedient.

Additional attention will need to be paid to how to decontaminate any facilities contaminated by a release. This may prove to be a bigger undertaking than dealing with the human exposure risks, as there is little experience in the literature on how to most cost effectively accomplish this task. Gases or liquids in aerosol form (e.g., formaldehyde) combined with surface disinfectants are often used to ensure complete decontamination. Gels and foams being pursued by scientists at Sandia National Laboratory (Zelikoff, 1998) can help in carrying and holding disinfectant to walls and ceilings. Curry and Clevenger (1997) recently reviewed promising research on biological decontamination by eight different "electro-technologies." These include electron beams, X-rays, pulsed electric fields, microwaves, and UV light. Of these, only UV light is likely to be feasible for patient decontamination, and then only with low-power UV in conjunction with a photosensitizer like hydrogen peroxide. Contaminated terrain often needs no decontamination other than natural drying and solar UV radiation, but exceptionally persistent organisms like anthrax need to be decontaminated using a spray mixture of chlorine-calcium, formalin, or lye solutions. In some locations seawater may serve as an expedient and less hazardous substitute (Manchee and Stewart, 1988).

Psychological Impact of Undergoing Decontamination

The psychological impact of being exposed to a poison is not well studied. Whether crowds will listen to instructions or panic, what they need to be told and how that message should be given, whether they will take off their clothes in the absence of an obvious immediate danger, whether they will shower with persons they have never met before, and how best to control or avoid hysteria are among the issues that need to be addressed.

MASS-CASUALTY TRIAGE PROCEDURES

The three primary objectives of a disaster response are: (1) do the greatest good for the greatest number of victims; (2) effectively utilize personnel, equipment, and health facilities; and (3) do not relocate the disaster from one location to another by poor command, control, or communication practices.

The triage process is the initial step taken to meet the primary objectives of a disaster response. The purpose of triage is to sort the injured by priority and determine the best use of available resources (e.g., personnel, equipment, medications, ambulances, and hospital beds). Many EMS agencies have in place a triage plan to implement in the event of an airplane crash, train derailment, or school bus accident. Traditional triage centers around the use of diagnosis-based criteria or involves the evaluation of each patient's respiration, perfusion, and mental status findings in order to determine whether they should be classified as urgent, delayed, or deceased. Both triage approaches require the examiner to see the patient and obtain certain clinical data by verbal communication and tactile examination. In a chemical terrorist incident the victim(s) may suffer from the effects of poison, trauma, or both. In a more conventional disaster, unless they are in danger, the patients can usually remain in place until directed to relocate. Their evacuation and treatment priority is indicated on a triage tag or colored ribbon. Unlike military triage protocols, where the focus is on successful completion of the "mission," the emphasis in the civilian sector is on saving as many persons as possible.

There are several differences between the triage done for the traditional disaster scenario and that for a hazardous material incident or a chemical/biological terrorist event. Time demands, patient volume, and the PPE being worn by response personnel in the hot and warm zones may preclude normal life-saving measures being rendered quickly, if at all. For example, verbal communication may not be possible because of the responder's PPE. A tactile examination may not be possible for the same reason. Additionally, the whole concept of traditional triage (treating the most seriously injured first) may not be applicable in a chemical or biological incident. Those walking around may need to be among the first to be decontaminated and evacuated because they have the best chance of survival. It is not desirable that victims remain in place in the hot zone until examined. Rather, immediate evacuation efforts should be undertaken and the victims directed towards the decon process established in the warm zone. Also, there will be little, if any, time to indicate a patient's priority on a triage tag in the hot or warm zones. Additionally, the patient data recorded on a triage tag is at risk of getting defaced when the tag becomes wet during decontamination.

Psychological issues also play a part in triage after a mass chemical or biological terrorist attack. Among the most important directions given to victims of nonhazmat incidents is how to evacuate the area, stop bleeding, and stay warm. The mixture of men with women and young and old together in this circumstance poses psychological problems.

R&D NEEDS

A comprehensive national training program on the medical management of patients injured by weapons of mass destruction (WMD) should be developed for prehospital and hospital personnel. The curriculum should include the following:

- site management/crowd control,
- triage,
- providing medical care while wearing PPE,
- set-up of mass decon areas in the field and at hospitals,
- performing mass decon on ambulatory and nonambulatory patients of all ages, and
- proper recognition and management of the psychological aspects of undergoing decontamination and exposure to WMD.

Little empirically based information exists in these areas, but it appears to the committee that equipment needs are secondary to information about procedures and methods.

7-1 The committee therefore recommends that research and development efforts in decontamination and mass triage be concentrated on operations research on procedures and techniques for effective decontamination of large numbers of people. Such research should include:

- *the physical layout, equipment, and supply requirements for performing mass decon for ambulatory and nonambulatory patients of all ages and health in the field and in the hospital;*
- *a standardized patient assessment and triage process for evaluating contaminated patients of all ages;*
- *optimal solution(s) for performing patient decon, including decon of mucous membranes and open wounds;*
- *the benefit vs. the risk of removing patient clothing;*
- *effectiveness of removing agent from clothing by a showering process;*
- *how much contact time for showering is necessary to remove a chemical agent;*
- *whether high pressure/low volume or low pressure/high volume spray is more effective for patient decontamination;*
- *the best methodology to employ in determining if a patient is "clean"; and*
- *the psychological impact of undergoing decontamination on all age groups.*

8

Availability, Safety, and Efficacy of Drugs and Other Therapies

This chapter reviews current and potential countermeasures for the chemical and biological agents. Discussion of chemical agents includes assessment of availability at both the first responder and local treatment facility level because of the need for rapid action in many cases. Treatment of victims of most of the biological agents being considered in this report is not so time dependent (in most instances there will not be any first responders involved), and discussion of availability therefore focuses on the existence and ease of purchase of required drugs and supportive equipment.

The discussions and the respective tables that follow permit a detailed analysis of these chemicals and biological agents. In a world of infinite resources almost all of the antidotal interventions being pursued for these potential agents would be of scientific merit. Many of these interventions might substantially advance scientific thought, and most could play a limited role in improving care, but all will confront the problem of Investigational New Drug (IND) status and FDA approval in the face of a very low natural incidence and ethical barriers to controlled testing in human subjects. These investigations will be exceptionally expensive and it is not apparent that the commercial pharmaceutical industry would consider research in this domain profitable without military or governmental support. In our world of finite resources a more pragmatic approach has been chosen for suggestions for research and development of antidotal agents. In particular, interventions that have a demonstrated benefit and might be improved upon are favored over novel approaches that have yet to be shown efficacious in human patients.

In addition, research and development recommendations are based on the premise that the most valuable treatments will be those that will be (or might be) useful even if a biological or chemical assault does not occur.

A third consideration in making recommendations was based on the committee's view that prophylactic interventions will rarely, if ever, be appropriate. The decision as to whether prophylactic therapy is appropriate for any of these biological or chemical agents must be based on several issues: risk to personnel, potential benefit for the individual and society, and extent of societal expenditure. The committee's view is that these considerations preclude any prophylactic interventions for the entire population, at least for the biological or chemical agents under consideration in this report. Certain prehospital first responders might be considered for prophylaxis against specific biological or chemical agents, but the scientific evidence in favor of prophylaxis of this smaller but still very large population is limited, and the risks and expenditures would still be substantial. In making this recommendation the committee's focus is purely civilian, and it should not be construed as discouraging the development of prophylactic interventions for use by the armed forces. The differences with regard to military and civilian prophylaxis strategies are substantial, encompassing not only the simple contrast of known threats at known times for military forces as opposed to unknown threats at unknown times for the civilian population, but also the levels of organization and systemic preparedness required and available.

Two general and important conclusions will become obvious to the reader as he or she proceeds through this chapter. The first is that with a few exceptions, drugs, antitoxins, and supportive medical equipment are generally available *in small quantities* (although two recent surveys [Dart et al., 1996; Skolfield, 1997] by poison control centers report that very few hospitals in their service areas carried sufficient amounts of all recommended antidotes). Proper planning and coordination among area medical and veterinary facilities might yield sufficient quantities of these drugs and other supplies for a multiple-victim incident, but few locales will have adequate supplies for a true mass-casualty event.

The second general conclusion is that many of the vaccines and therapeutics described below are only available under Investigational New Drug (IND) applications to the Food and Drug Administration (FDA). Such products are generally produced in limited amounts and can be used only in a research setting and with the informed consent of the recipient (i.e., the patient or a proxy must provide informed consent, and the FDA must be contacted for an IND number for the patient before the manufacturer can provide the product). In some cases, a fully licensed FDA-approved product will emerge after the requisite evidence of safety and efficacy is accumulated. In the interim however, under current legal requirements, IND status will effectively preclude use in a mass-casualty situation. Furthermore, it will be difficult or impossible to collect the required evidence of efficacy for many INDs (randomized clinical trials in human patients), either because the disease is so rare that accumulating enough cases will take a very long time, or because the condition against which it is directed does not occur naturally (e.g., mustard poisoning). Earlier this year, FDA established rules making it easier to study investigational drugs and devices with patients in life-threatening situations and unable to give informed consent. However, these rules, which require extensive prior planning, are aimed at facilitating collection of efficacy data and do not directly address the mass-casualty situation, especially for terrorist acts involving chemical and biological agents.

FDA recognized the difficulty IND status presented in potential mass-casualty situations during the Persian Gulf War and passed an interim rule waiving the requirement for the United States military to obtain informed consent in using two investigational products intended to provide protection against chemical and biological warfare agents (pyridostigmine bromide and botulinum toxoid vaccine). The FDA has recently solicited comments on the wisdom of revoking this interim rule as well as on the nature of the evidence that ought to be required when products cannot ethically be tested in humans (United States Food and Drug Administration, 1997).

CHEMICAL AGENTS

Discussion of chemical agents is based upon an approach that integrates local, state, and federal systems for the delivery and stockpiling of antidotes for mass casualty events. This approach emphasizes which agents must be available locally, how much and under whose jurisdiction. The principle being that a plan should be developed to deliver large quantities of antidotes to any part of our country in a rapid organized fashion. Research and development should focus on models of storage and methods for deployment and delivery in a timely fashion. First responders from Emergency Medical Services and Hazmat Services cannot be expected to make definitive decisions and in general will not be stocked for population antidotal care, although they should have access to personal antidotal material for high-risk toxins so as to effectively complete scene assessment and victim rescue.

Nerve Agents

The treatment for nerve agent poisoning recommended by the U.S. military involves the use of three therapeutic drugs: atropine, pralidoxime, and diazepam. Nerve agents act by binding to the enzyme acetylcholinesterase, thereby blocking its normal function of breaking down the neurotransmitter acetylcholine following its release at neuronal synapses and neuromuscular junctions throughout the peripheral and central nervous systems. Acetylcholine accumulates and overstimulates synapses throughout the brain, nervous system, glands, and skeletal and smooth muscles. Death is usually caused by respiratory failure resulting from paralysis of the diaphragm and intercostal muscles, depression of the brain respiratory center, bronchospasm, and excessive bronchial secretions. Seizure activity also contributes to morbidity and mortality.

Atropine sulfate is a drug that blocks muscarinic acetylcholine receptors, counteracting effects such as vomiting and diarrhea, excessive salivation and bronchial secretions, sweating, and bronchospasm. It is administered intravenously, if possible, in high doses at frequent intervals until signs of intoxication diminish. Pralidoxime chloride (2-PAM), a drug that reactivates the nerve agent-inhibited cholinesterase, is administered along with atropine. Diazepam, or another anticonvulsant, may be administered in severe cases to control seizures and thereby prevent seizure-induced brain damage.

Appropriate adult doses of atropine sulfate, 2-PAM, and diazepam are packaged in autoinjectors issued to U.S. military personnel for self- or buddy-aid. A metered dose atropine methonitrate inhaler called the medical aerosolized nerve agent antidote (MANAA) has been approved by FDA and is being produced for DoD by 3M/Riker. However, it is intended for use, under medical supervision, as a supplement to injectable atropine, not as self/buddy aid. Except under special circumstances, utilization of these prepackaged autoinjectors should be limited to Hazmat and prehospital EMS staff for their own personal care and that of their coworkers. Consideration for use of these antidotes for the general public should be restricted to exceptional circumstances when patients cannot be expeditiously removed from the environment, decontaminated, and brought to an emergency department. None of these antidotes is ideally delivered intramuscularly, in the absence of intravenous fluids and control of the airway, or during a convulsion. If these are considered essential products for civilian care, the hospital emergency department is the ideal site for their use.

In a nerve agent incident where a presumed exposed patient is to be decontaminated prior to transportation to an emergency department, it can be considered appropriate for prehospital medical personnel to utilize prepackaged antidotes (atropine sulfate, diazepam and pralidoxime chloride) if and only if:

1. There are signs and symptoms indicative of nerve agent poisoning, namely, meiosis, rhinorrhea, shortness of breath, fasciculations, or seizures.
2. There is an initial intelligence basis for suspecting the presence of a nerve agent at the scene or a high quality detection system that indicates the presence of a nerve agent at the scene.
3. A qualified physician with skills in medical toxicology is actively involved in the management of the patient.
4. The antidotes are utilized before or during decontamination and in no way delay transfer to a health care facility or casualty collection point.

If transfer to a health care facility subsequent to decontamination will exceed 30 minutes, it may be appropriate to treat additional civilians at the scene. The committee is aware of no studies performed comparing central nervous system levels and benefits achieved by intravenous administration of these antidotes with those achieved by intramuscular injection performed 15–45 minutes earlier. Such a comparison would be an important consideration in deciding upon expedient prehospital treatment.

An alternative to extensive field treatment by Hazmat, EMS, and MMST teams in a particular region might utilize Hazmat and MMST teams as a mobile stockpile system delivering large quantities of antidotes to the EMS teams/ambulances (and individual hospitals as patients move there). This approach will ensure that patient load at a given hospital will be matched by antidote supply, thus expediting therapy and avoiding delays in delivery from a single central stockpile. Decisions on antidote stockpiling and control will involve geographic (rural vs. urban), financial, and other legitimate but nonscientific determinations, but in the proposed procedure, first responders would draw on established supplies of antidotes prepared for disaster management to ensure that patients transported to local emergency departments arrived with sufficient antidotes to begin treatment. Simultaneous communication with Regional Poison Control Centers and Poison Control Center–Emergency Department linkages to local and state health departments would track stockpile usage and allow for coordination with more distant sources, such as the Centers for Disease Control and Prevention.

In nonhuman primate studies, the combination of atropine and 2-PAM will protect against up to five times the LD_{50} (the dose lethal to 50 percent of the population exposed)[1] of all known nerve agents except

---

[1]LD_{50} is a statistical concept rather than a clinical one, so neither doses below the LD_{50} nor protection against doses even higher than the LD_{50} guarantee that everyone exposed will survive.

soman (GD). Soman is an exception, because 2-PAM acts by competitively binding to the organophosphate agent and thereby "reactivating" the acetylcholinesterase enzyme the agent had tied up. However, once the enzyme-agent complex has undergone an irreversible "aging" process, 2-PAM is unable to reactivate the enzyme. The aging process takes hours for VX and most of the G agents, but only minutes in the case of soman (GD). In most cases of domestic civilian terrorism involving soman intoxication, it will not be possible to administer 2-PAM this quickly. Additional limitations in the use of 2-PAM as an antidote in nerve agent toxicity include the fact that large doses may be necessary for protection and survival, but in such large doses 2-PAM itself can lead to significant side effects, most notably hypertension. In addition, because it does not readily cross the blood-brain barrier, 2-PAM is thought to have little action against the central nervous system effects of nerve-agent poisoning.

Although 2-PAM and atropine sulfate have only limited efficacy against soman (GD), nonhuman primates given the peripherally acting carbamate pyridostigmine prior to exposure to the nerve agent and atropine sulfate and 2-PAM after exposure survived GD in doses up to 20 to 40 times the LD_{50}. Pyridostigmine appears to be without comparable benefit in treatment of sarin or VX, however. Like the nerve agents, carbamates inhibit the enzymatic activity of acetylcholinesterase. In fact, carbamate-enzyme binding precludes organophosphate-enzyme binding. Unlike the nerve agents, however, the carbamate-enzyme bond is freely and spontaneously reversible. As a result, it is possible to protect acetylcholinesterase from irreversible inhibition by nerve agent by use of the reversible carbamate inhibitor. The use of pyridostigmine by large numbers of military personnel for periods of 6–7 days during the Gulf War resulted in uncomfortable but not disabling side effects (primarily gastrointestinal and urinary) in more than half of those taking the drug (Dunn et al., 1997). In most cases these effects subsided after a day or two. Numerous controlled studies in humans, as well as years of use in the treatment of myasthenia gravis, support claims for the safety of pyridostigmine.

The utilization of prepackaged diazepam for intramuscular use is a poor parenteral therapeutic delivery technique for this anticonvulsant. The diazepam is dissolved in propylene glycol and is poorly and erratically absorbed following intramuscular use. Although the intramuscular route is considered to be the least effective route for seizure control, lorazepam can be used intramuscularly and could be preferred to diazepam for EMS and Hazmat use. Lorazepam, however, has several disadvantages. From a financial perspective it is more expensive than diazepam. Lorazepam is not stable at high temperatures and therefore cannot be as easily stored as diazepam. Finally, without preloaded syringes or autoinjector packaging, intramuscular use will be difficult to accomplish efficiently while utilizing the protective clothing required at the scene.

Organophosphate (OP) pesticides are widely used throughout the United States, and poisoning is common (Litovitz et al., 1997). Treatment is identical to that for nerve agents, and as a result, many emergency medical teams and most hospital emergency department staff have some familiarity with diagnosis and treatment of OP poisoning and have access to limited supplies of atropine and pralidoxime. However, multiple nerve agent victims may each need 10–50 milligrams (mg) of atropine sulfate, which would rapidly deplete supplies in receiving hospitals. Rural communities may be able to call on veterinarians, who sometimes hold substantial amounts of atropine to treat cattle or horses poisoned by organophosphate pesticides. They might also be sources of other drugs, resuscitation equipment. disinfectants, and other useful equipment and supplies (Schneider, 1987). The same general concern—treatment would be possible only for small numbers of patients—is also true with regard to availability of ventilators. As in many other disaster situations, intubated patients can be supported by bag valve mask ventilation until a ventilator is available. Bronchoconstriction and copious secretions are prominent effects of organophosphate poisoning, and therefore ventilation is likely to be required for up to several hours after exposure, even when appropriate drug therapy is available.

Potential Advances

Table 8-1 provides information on a number of treatments and prophylactic pretreatments in various stages of research and development. This table and those that follow contain the relativistic term "potential civilian utility" and employ a very liberal criterion in assessing products for such use. The accompanying text evaluates potential products in a more selective manner that emphasizes probability and priority. For example, various pralidoxime derivatives, such as Pro-2-PAM, P-2-S and the Hagedorn oximes such as HI-6, have been compared to 2-PAM. Although some of these products offer increases in efficacy under some circumstances, none are FDA approved and most have intrinsic formulation and stability problems. The committee recommends that no further investment be made in attempting to bring these or similar compounds to market and/or to establish stockpiles The potential cost appears far more substantial than the advantage they might provide over 2-PAM.

Alternatives to atropine sulfate autoinjectors, such as the quaterary ammonium derivatives ipratropium bromide and atropine methonitrate, have the disadvantage of poor absorption across mucosae and the blood-brain barrier, resulting in prolonged local effects, but they have negligible

TABLE 8-1 Potential Antidotes for Nerve Agent Poisoning

Antidote	Efficacy	Availability	Potential Civilian Utility	Stockpile
Scopolamine	Poorly absorbed through inhalation	Yes	Yes	Hospitals, ED
Ipratropium Bromide[a]	A quaternary anticholinergic agent	Yes for inhalation	Yes	Hospitals, ED
3-quinuclidinyl benzilate (BZ, QNB)[b]	CNS effects	Withdrawn	None	N/A
N, N′ trimethylene bis (pyridine-4-aldoxime bromide) combined with benzactyzine	CNS effects Benzactyzine is a cholinolytic agent	Benzactyzine withdrawn	None	N/A
Benzodiazepines (Diazepam, Lorazepam, Midazolam)[c,d]	Controls seizures	Autoinjector: 10 mg/2-ml vials (convulsant antidote)	Field: yes Lorazepam preferred Base: no Intravenous preferred	Prehospital, Emergency, Hospitals Health Department
Pro-2-PAM (Dihydropyridine derivative)	Prodrug or drug carrier permits traversing blood brain barrier	Experimental	Inadequate evidence	N/A
Obidoxime (Toxogonin)	Effective in rodent model	European Countries	Inadequate evidence	N/A

Continued on next page

TABLE 8-1 *Continued*

Antidote	Efficacy	Availability	Potential Civilian Utility	Stockpile
H Series of oximes TMB4 (Hagedorn) (HI-6 compounds)[e-h]	Effective in rodent model Toxicity profile under study Direct central and peripheral anticholinergic activity	Stability in question	Inadequate evidence	N/A
Methanesulfonate salt of pralidoxime (P2S)		Standard in UK	Inadequate evidence	N/A
Nicotine hydroxamic acid methiodide (NHA)	Pretreatment of soman exposure in rhesus monkeys	Research potential		N/A
Monoisonitrosoacetone (MINA)	May have 48 hr post-exposure utility	Research potential		N/A
Butyryl cholinesterase (BChE)[i-k] (Human BChE Mutants)	Exogenous scavenger for highly toxic organophosphorus poisons. Equine BChE studied in rhesus monkeys	Human BChE available (? FDA status)	Prehospital: high-risk environment pretreatment essential personnel ED: no	N/A

Agent	Description	Development status	Indication	ED
Stoichiometric scavengers: acetylcholinesterase[l,m] and carboxylesterase[n,o] [fetal bovine products]	Binds broad spectrum of nerve agents. Longer half-life. Pretreatment common disadvantage that they have high MW and react 1:1 with organophosphates. Effective at 1:1 concentration.	Research	Prehospital: high-risk environment pretreatment essential personnel ED: no	N/A
Catalytic scavengers: Organophosphorus acid anhydride hydrolase,[p,q] [parathionase] modified AChE, modified BChE	Pretreatment: advantage small amount of effective enzyme to destroy large amounts of toxin.	Research	No	N/A
Catalytic monoclonal antibodies[r,s]	Mice hybridomas secret monoclonal antibodies which hydrolyze phosphonates. The antibody is an IgG_{2a} with Kappa light chain character with activity against soman, but there is no cross reactivity against sarin or tabun.	Rodent models		N/A
Reactive topical skin protectants	Protection against penetration and will detoxify nerve agents.	Animal model	Prehospital and ED personnel. Field use in high-probability zone	N/A

Continued on next page

TABLE 8-1 Continued

Antidote	Efficacy	Availability	Potential Civilian Utility	Stockpile
Memantine[t]	May be neuroprotective in cell culture for soman but severe injury still noted.	Antiparkinsonian agent	No evidence	N/A
Thienylcyclohexylpiperidine (TCP)[u]	Acts as a noncompetitive inhibitor of NMDA receptors. May prevent and interrupt soman-induced seizures. Better approaches available.	Experimental	No evidence	N/A
Dizocilpine (MK-801)[v]	Act as a noncompetitive inhibitor of (N-Methyl-D-Aspartate) NMDA receptor channel during seizures induced by soman in guinea pigs. Adverse effects similar to those of PCP.	Experimental	No evidence	N/A

[a]Gross, 1988; [b]Waelbroeck et al., 1991; [c]Martin et al., 1985; [d]McDonough et al., 1989; [e]Lundy et al., 1992; [f]Koplovitz and Stewart, 1994; [g]Worek et al., 1995; [h]Kassa, 1995; [i]Raveh et al., 1993; [j]Broomfield et al., 1991; [k]Masson et al., 1993; [l]Maxwell et al., 1991; [m]Velan et al., 1992; [n]Maxwell et al., 1987; [o]Doctor et al., 1993; [p]Little et al., 1989; [q]Ray et al., 1988; [r]Lenz et al., 1992; [s]Lenz et al., 1984; [t]Deshpande et al., 1995; [u]Carpentier et al., 1994; [v]Sparenborg et al., 1992.

systemic effects. Further comparison of inhaled atropine methonitrate, scopolamine, and ipratropium bromide with intramuscular atropine is indicated.

Of the diverse agents with potential as catalytic and stoichiometric scavengers, most remain in the early research stages and, as is the case with other pretreatments, their potential utility in managing the consequences of a domestic civilian terrorist incident involving nerve agent is not clear. Further development of human butyrlcholinesterases by the DoD could provide a potential pretreatment for Hazmat and prehospital staff performing rescue in unsafe environments. The relationship between effectiveness and time of delivery relative to nerve agent exposure is essential in evaluating that possibility. Only exceptional intelligence and information-sharing will provide these first responders with the time likely to be necessary for effective use of these scavengers or any pretreatments requiring substantial lead times.

R&D Needs

8-1 *Atropine sulfate, pralidoxime, and diazepam autoinjectors and stockpiles of these drugs should be available both for onsite self/buddy use by emergency medical personnel and for delivery to hospitals or patient collection points with patients. In addition atropine, pralidoxime and diazepam should be readily available in large quantities from a stockpile controlled by the local health department to be brought to the site where EMS will bring the casualties. Studies of stockpile control and time necessary for the delivery to prehospital, hospital and health departments should be performed for each region. Specifically, a study should be designed to describe the most effective distribution system for a mass casualty event. These studies must emphasize an integrated analysis based on the potential of regional health, police, and fire personnel.*

8-2 *A comparison of the central nervous system levels and benefits achieved by intravenous administration of atropine and 2-PAM with those achieved by intramuscular injection performed 15–45 minutes earlier would provide important help in deciding upon the criteria for and amount of prehospital treatment to recommend.*

8-3 *Lorazepam should be investigated as an intramuscular anticonvulsant for use in the field by EMS and Hazmat personnel. A study of its stability at room temperature will be essential for its use in the field.*

8-4 *Needle, intravenous, and/or intravenous bag adapters should be designed to facilitate the intravenous delivery of antidotes currently prepared solely for intramuscular use. The prepackaged systems designed for intramuscular use can only be used intravenously with risk to the health professional giving the injections.*

8-5 Further comparison of inhaled atropine methonitrate, scopolamine, and ipratropium bromide with intramuscular atropine is indicated.

8-6 New more effective anticonvulsants are needed for autoinjector applications. Anticonvulsants that are water soluble and effective in halting nerve
agent-induced seizure activity as well as preventing recurrence of the seizure
activity will improve the recovery of seriously poisoned casualties.

8-7 The committee recommends support of continuing research to develop catalytic scavenger molecules such as human butyrylcholinesterase and carboxyes-
terase as both potential pretreatments against anticholinesterases and as immediate postexposure therapies.

8-8 Research into the development of catalytic monoclonal antibodies against a
broad spectrum of nerve agents may prove beneficial in the development of
rapid diagnostic tests as well as in the development of potential new therapies.

Vesicants

Included in this category of chemical agents are various forms of
"mustard," an arsenical compound called Lewisite, and phosgene oxime.
No evidence suggests that Lewisite or phosgene oxime has ever been
used on the battlefield, but sulfur mustard (bis [2-chloroethyl] sulfide) has
been used in several wars, most recently in the Iran-Iraq conflict, and it is
considered the most likely to be used on the battlefield. An immediate
precursor, thioglycol, has many industrial uses and is available commercially. A simple substitution reaction yields mustard. Sidell et al. (1997) is
the primary source of the information presented in this section.

The name mustard apparently stems from the compound's smell,
taste, and color rather than any chemical resemblance to the popular
spice. At room temperature sulfur mustard is an oily liquid that is only
slightly soluble in water and therefore very persistent in the environment.
At higher temperatures it becomes a significant vapor hazard ("mustard
gas"). It quickly permeates rubber and is readily absorbed by skin. eyes,
airway, and gastrointestinal (GI) tract. It reacts within minutes with components of DNA, RNA, and proteins, severely compromising normal cell
function. Acute local effects can be severe enough to require days to weeks
of care, but mortality, usually from pulmonary insufficiency or superim-
posed infection, is low. No effective treatment of mustard damaged tissue
is currently available. Immediate decontamination of exposed skin areas
is the only means of preventing tissue damage from mustard. U.S. military publications recommend 0.5 percent sodium hypochlorite followed
by soap and water, or the resin-based M291 and M295 decontamination
kits. This task is made more difficult by the fact that clinical signs, including pain, are not evident for 2–12 hours, depending on the dose and tissue
exposed. Eyes may be flushed with copious amounts of water. Skin, eye,

and airway damage is treated similarly to thermal burns, and pain relief is provided by topical or systemic analgesics (Willems, 1991). Early intubation and oxygen therapy are recommended for patients with signs of airway damage.

Lewisite (β-chlorovinyldichloroarsine) was synthesized in 1918 for use as a weapon, and its clinical effects are similar to those of mustard in many respects, although the cellular mechanisms are believed to differ. However, unlike mustard, Lewisite liquid or vapor produces irritation and pain seconds to minutes after contact. Immediate decontamination may limit damage to skin or eyes, and intramuscular injections of a specific antidote, dimercaprol, or British antiLewisite (BAL) will reduce the severity of systemic effects. BAL has toxic effects of its own, however, and must be used with care.

Phosgene oxime (dichloroformoxime) is a colorless crystalline solid with a melting point of approximately 37.7°C (100°F). In liquid or vapor form it is highly corrosive, and it penetrates clothing and rubber readily. The mechanism by which it damages tissue is unknown, but its effects are almost instantaneous and produce severe pain. Skin lesions are like those caused by a strong acid. There is no antidote; treatment will be similar to that for mustard.

Table 8-2 provides information about ongoing work on potential countermeasures in various stages of development. All the entries are drawn from current work by DoD labs, primarily USAMRICD. Considerable basic research is devoted to better delineation of the mechanism(s) of action in order to develop protective and ameliorative interventions. Strategy to date has involved parallel investigation of intracellular scavengers, cell cycle inhibitors, calcium modulators, protease inhibitors, and antiinflammatory drugs. The rapid action of the vesicants, the lack of immediate pain in the case of mustard, and the attractiveness of an attack employing the vapor (ergo, pulmonary) route make decontamination an unsatisfactory strategy for civilians, and, as is the case with most of the agents being considered, pretreatments requiring any more than a few minutes lead time are not likely to be generally useful in coping with civilian terrorism incidents. Topically applied skin protectants offer the possibility of protection from trace amounts of agent-penetrating protective garments or surviving decontamination of equipment, but certainly could not be counted on to replace chemical-resistant clothing in areas known to be contaminated. The current research aimed at moderating or repairing vesicant injury is therefore extremely important, despite, or perhaps due to, the fact that most of the candidate drugs are still years away from licensing. Clinical testing of efficacy in humans is not possible, so early agreement with FDA on surrogate measures will be critical.

TABLE 8-2 Potential Additional Countermeasures for Vesicant Agent Poisoning

Antidote	Efficacy	Availability	Potential Civilian Utility	Stockpile
Topical skin protectant (TSP)	Passive protection	Goal is FDA license by FY00	Prehospital high-risk personnel	Health Dept.
Reactive TSP (decontaminates)	Proof of principle	Goal is FDA license by FY08	Prehospital high-risk personnel	Health Dept.
Nitric oxide synthase inhibitor[a] Nitroarginine methylester (NAME) effective at high concentrations	Cell culture vs. sulphur mustard	Preclinical	Insufficient evidence	N/A
Combinations of dexamethasone, heparin promethazine, vitamin E, sodium thiosulfate[b]	Preliminary evidence in rats	FDA-approved drugs	Insufficient evidence	N/A
Sodium thiosulphate (i.v.)	Animal data shows effects up to 20 min post exposure	FDA-approved drug	Insufficient evidence	N/A
N-acetyl Cysteine	Protection from vapors in rats	Preclinical	Insufficient evidence	N/A
Calcium chelators	Protects skin cells in culture	Preclinical	Insufficient evidence	N/A
CO_2 laser debridement	Mustard ion weanling pigskin	Preclinical	Insufficient evidence	N/A

[a]Sawyer et al., 1996; [b]Vojvodic et al., 1985.

R&D Needs

> *8-9 The lack of treatments for vesicant injury constitutes a serious shortfall in civilian medical preparedness, and the existing program of research into mechanism should be supplemented by an aggressive screening program focused on repairing or limiting vesicant injuries, especially airway injuries.*

Cyanide

The cyanide anion, CN⁻, whether delivered in hydrocyanic acid or in a cyanogen such as cyanogen chloride, exerts its toxicity primarily by inhibiting mitochondrial cytochrome oxidase, which leads to lactic acidosis, cytotoxic hypoxia, seizures, dysrhythmias, respiratory failure, and death within minutes after inhalation or oral ingestion of a large dose (1 to 3 mg/kg of body weight). One antidote for cyanide poisoning is amyl nitrite, which converts hemoglobin to methemoglobin, which in turn competes effectively for cyanide with the mitochondrial cytochrome oxidase complex. Intravenous sodium nitrite is generally used for this purpose after an initial dose of the volatile amyl nitrite is given by inhalation. Cyanide is then removed from cyanomethemoglobin by intravenous sodium thiosulfate, which reacts with cyanide to form nontoxic thiocyanate. Gastric lavage with activated charcoal should be administered if cyanide is ingested. Supportive therapy includes intubation, correcting acidosis, and, if necessary, administering anticonvulsants. Cyanide is metabolized more readily than the other chemical agents, and as a result, if the initial dose is not so large as to kill the victim within minutes, supportive therapy may be sufficient for full recovery in a matter of hours.

Amyl nitrite, sodium nitrite, and sodium thiosulfate are commercially available in standard doses in the Pasadena Cyanide Antidote Kit (formerly the Eli Lilly Cyanide Antidote Kit). Many poison control centers and emergency departments may have small quantities of such kits on hand. As in the case of nerve agents, a mass-casualty situation will quickly exhaust supplies. Pooling resources from the whole community could be beneficial, but only if communications and a mechanism for sharing pre-incident intelligence are already in place. As suggested above, the use of prehospital medical personnel to ensure that antidotes are delivered to the hospitals and/or casualty collection points with the affected patients is a potential means of matching antidote supply with patient load. As in the case of nerve agent incidents, the committee recommends that emergency medical teams responding to an event bring a substantial number of these antidote kits and move the kits with the patients to assure adequate therapy is available on emergency-department arrival. Only if substantial delay is expected due to unavailability of a local ED or if cyanide has been conclusively identified as the toxic substance involved should consideration be given to prehospital use of this antidote kit. The use of the cyanide antidote kit by prehospital personnel under uncertain circumstances would place civilians at risk for substantial delay in transport to a health care facility. This antidote kit is a three-part intervention requiring experienced clinical judgment prior to use and demanding intravenous access for the major components of therapy. Use of the kit demands substantial time, particularly for a heterogeneous population including children, the aged, and medically compromised individuals.

Potential Advances

Additional interventions not yet licensed or available for use are summarized in Table 8-3. Although a number of entries in the table have shown great promise in preclinical studies, pretreatment is not a viable option in the most probable civilian terrorism scenarios. The 8-aminoquiniline compound WR242511, for example, is reported to protect mice from a dose of cyanide 5 times the LD_{50}, but it must be administered 8 hours before cyanide exposure. The initial three entries therefore rate a higher priority than the remainder, which may eventually provide a measure of protection for military troops, cyanide industry workers, or first responders in the vicinity of large stores of cyanide, but will not be effective post exposure. Three postexposure treatments are currently approved drugs in Europe and might be licensed in the United States without great delay if there were a perceived market for them. The advantage of these newer drugs is their diminished acute risk/benefit in children and the physically compromised patients. On the other hand, the major difficulty with using them for current treatment protocol is the inability to treat the victim within minutes of exposure, a problem that will not be remedied by new drugs. In general, if the patient has survived to the time of arrival of clinical support, the probability of survival is great.

R&D Needs

> 8-10 *The organization of delivery and availability of adequate supplies of the cyanide antidote kit must be achieved. Studies of stockpile control and time necessary for the delivery to prehospital, hospital, and health departments should be performed for each region. Specifically, a study should be designed to describe the most effective response system for a mass-casualty event. These studies must emphasize an integrated analysis based on the potential of regional health, police, and fire personnel.*

8-11 Further understanding of the risks and benefits of methemoglobin forming agents should be investigated.

8-12 A continued investigation into the benefits of hydroxocobalamin and stroma-free methemoglobin would be valuable and is an appropriate avenue of investigation.

8-13 Dicobalt ethylene diamine tetraacetic acid and the strong methemoglobin forming compounds 4-dimethylaminophenol and various aminophenones merit further investigation, but must be given a lower priority then hydroxocobalamin and stroma-free methemoglobin, which carry less risk of creating excessive and unpredictable levels of methemoglobin.

Phosgene

Although phosgene is not currently believed to be a significant threat as a military weapon, it was used in World War I artillery shells, and, more importantly to the present discussion, it is still a widely used industrial chemical, over a billion pounds of which are produced in the United States. It is generally stored and transported as a liquid, but its low boiling point (7.5°C) means that it readily becomes a heavier-than-air gas. Pulmonary edema is the most serious consequence of inhalation—onset within 2 to 6 hr. is indicative of severe injury. Low concentrations may produce mild coughing, dyspnea, and a feeling of discomfort in the chest, although individuals may remain essentially asymptomatic for up to 72 hours. Physical exertion during this period may precipitate signs and symptoms. Rest is thus an essential component of patient management, along with airway management (control of secretions and bronchospasm) and oxygen therapy. There are no proven pharmacological interventions for pulmonary phosgene at present, so, as with the vesicants, rapid removal from the source and thorough decontamination is essential. Table 8-4 provides information on other treatments that have been or are being investigated.

R&D Needs

8-14 Protection of the pulmonary bronchi and bronchioles may be possible by the use of cytoprotective agents. Ongoing studies of pneumocytes, interstitial, and epithelial cells suggest that antiinflammatory agents, such as aminophylline, corticosteroids, and ibuprofen, may be useful. Further studies on the ability of N-acetylcysteine to limit the inflammatory cascade produced by effects of phosgene and its metabolic byproducts are also justified, as are additional studies of its systemic antioxidant effects.

TABLE 8-3 Potential Additional Antidotes for Cyanide

Antidote	Efficacy	Availability	Potential Civilian Utility	Stockpile
Hydroxocobalamin (Vitamin B_{12a})[a-e] (+NaHSO$_4$) Kit	No methemoglobin Low toxicity High CN affinity	Short shelf life (France) Orphan drug	Post exposure	Health dept. emergency dept.
Dicobalt ethylene diamine tetraacetic acid (EDTA) (Kelocyanor)[f]	IV Risk: Cardiac Dysrhythmias angina, death	Europe: commercial USA: Experimental	Post exposure	N/A
4-Dimethylaminophenol (4-DMAP)[g,h] and similar molecules P-aminopropiophenone (PAPP),[i] P-aminoheptanophenone (PAHP), P-aminooctanoylphenone (PAOP)[j]	IV, IM Possible Mutagen Local tissue necrosis Marked ↑ methemoglobin Temperature ↑, pain PAHP (safest?)	Germany	Post exposure	N/A
Stroma free methemoglobin[k-m]	Experimental	No	Postexposure and Prehospital high-risk personnel	Health dept. Poison center

Superactivated charcoal[n,o]	For oral exposure	FDA approved	Postexposure and Prehospital high-risk personnel	Emergency dept.
α-adrenergic antagonists (chlorpromazine;[p] phenoxybenzamine)	Mechanisms uncertain	FDA-approved drugs		N/A
8-aminoquinoline analogs of primaquine[q] (e.g., WR242511)	Methemoglobin formers Pretreatment	Varies with compound	Prehospital high-risk personnel	N/A
Alpha-ketoglutaric acid[r–u]	Direct binding of cyanide without methemoglobin formation Animal studies only	Experimental	Insufficient evidence	N/A

[a]Hall and Rumack, 1987; [b]Cottrell et al., 1978; [c]Brouard et al., 1987; [d]Bismuth et al., 1988; [e]Beregri et al., 1991; [f]Hillman et al., 1974; [g]Weger, 1983; [h]Bhattacharya, 1995; [i]Marrs and Bright, 1987; [j]Rockwood et al., 1992; [k]Ten Eyck et al., 1983; [l]Ten Eyck et al., 1986; [m]Breen et al., 1996; [n]Andersen, 1946; [o]Lambert et al., 1988; [p]Peterson and Cohen 1985; [q]Steinhaus et al., 1990; [r]Dulaney et al., 1991; [s]Bhattacharya and Vijayaraghavan, 1991; [t]Hume et al., 1995; [u]Norris et al., 1990.

TABLE 8-4 Potential Antidotes for Pulmonary Phosgene

Antidote	Efficacy	Availability	Potential Civilian Utility	Stockpile
Hexamethylene tetramine[a] (methenamine; urotropin, HMT)	Limited prophylactic effect; no convincing benefit after exposure	Yes	None	N/A
Cysteine[b,c]	Traps phosgene and converts to less harmful metabolites.	Yes	Insufficient evidence	Hospital emergency dept.
N-acetylcysteine[d]	Evidence in rabbits only	Yes	Prehospital? Pre- and post-exposure	Hospital emergency dept.
Corticosteroids[e]	To decrease inflammation Limited evidence	Yes	Yes	Hospital emergency dept.
Aminophylline[f]	Preexposure and postexposure Attenuates lipid peroxidation	Yes	Yes	Hospital emergency dept.

[a]Diller, 1980; [b]Bridgeman et al., 1991; [c]Lailey et al., 1991; [d]Scuito et al., 1995; [e]Lorin and Kulling, 1986; [f]Sciuto et al., 1997.

BIOLOGICAL AGENTS

The following section on biological agents begins with a review of vaccine research followed by reviews of bacterial infections, rickettsia, viruses, and toxins that include agent specific R&D needs. The section concludes with nonspecific defenses against biological agents that show promise for the future.

Vaccines

Vaccines are the cheapest and most effective defense against a large number of infectious diseases. Public health vaccine programs are the principal means of providing protection to at-risk populations against a growing list of natural infectious disease hazards. As vaccine technology continues to dramatically improve and new vaccines are developed and licensed, the list of vaccine-preventable diseases is increasing. Laboratory workers are provided protection against several highly hazardous bacteria and viruses which are considered to be potential biologic weapons through the use of vaccines in IND status under open protocols. Military populations can now be protected against the hostile use of several biologic threats by vaccination. The armed services have two licensed vaccines suitable for routine use if needed and have several more vaccines under development for protection of military populations against biologic threat agents (see Table 8-5).

Vaccination has limited value as a primary defense for civilian populations for several compelling reasons. The risk of exposure to a biologic threat agents is very low and uncertain for the general population. Pre-exposure vaccination of an entire population is a huge and daunting task. Achieving a high level of vaccine coverage of the U.S. adult population has never been done and probably could not be done in the absence of an eminent and credible threat. The costs and risks of vaccination are far too great and far outweigh potential benefits in view of the current assessment of the potential threat. Finally, there is a long list of potential agents and only a few licensed vaccines. The spectrum of achievable protection at present includes only smallpox, anthrax, and plague.

Vaccines against biologic threat agents do, however, have some very important uses in the civilian response to the threat of biologic terrorism. Anthrax vaccine can be effectively used in conjunction with antibiotics to prevent the development of pulmonary anthrax in exposed individuals. Botulinum toxoid vaccines can be used to immunize plasma donors to produce specific immune globulins for therapy of botulism. Smallpox vaccine (vaccinia) would be essential to prevent further spread of the disease following diagnosis of the originally exposed individuals. Vac-

TABLE 8-5 Vaccines Against Biologic Threat Agents

Agent or Disease	Stage of Development	Type	Civilian Utility
Anthrax	licensed	inactivated toxins	yes
Plague	licensed	inactivated bacteria	no
Tularemia	IND	live attenuated	no
EEE	IND	inactivated virus	no
WEE	IND	inactivated virus	no
VEE	IND	live attenuated	no
VEE	IND	inactivated virus	no
Botulism	IND	toxoids	yes
Q Fever	IND	inactive antigen	no
Smallpox	licensed	avirulent vaccinia virus	yes
Smallpox	IND	avirulent vaccinia virus	yes
Ebola/Marburg	preclinical	viral replicon/DNA	no
SEB	preclinical	toxoid	no
Ricin	preclinical	toxoid	no
Brucellosis	preclinical		no
Yellow Fever	licensed	live attenuated virus	no
Rift Valley fever	IND	inactivated virus	no
Rift Valley fever	IND	live attenuated virus	no
Junin virus	IND	live attenuated virus	no
Hantaan virus	IND	engineered vaccinia	no
Dengue	IND	live attenuated virus	no

cines against several of the threat agents may be used to immunize personnel in public health and research laboratories who must work with live agents. Several vaccines in IND status have been used with FDA approval in at-risk laboratory personnel for many years with appropriate informed consent and a suitable open protocol. Consideration might also be given to immunization of response team personnel and selected medical personnel in high-risk areas such as autopsy suites and microbiology laboratories. Consideration should also be given to immunization of selected law enforcement and intelligence personnel in high-risk assignments such as the White House or one of the national-level rapid response teams. In both of these cases, high risk will be a highly subjective judgment, given the lead time of weeks to months required for immunization to achieve its full effect.

The value of vaccines, even in the limited uses described above, justifies an accelerated research effort to improve the licensed vaccines, such as the anthrax, plague, and smallpox vaccines, and to complete the development process and seek licensure for those vaccines still in IND status.

The DoD has an ongoing research program aimed at improving or completing the development of the vaccines on the above list. A major DoD procurement effort, the Joint Vaccine Acquisition Program, has awarded a contract for the development and manufacture of many of the vaccines in Table 8-5. Completion times to fulfill the contract call for most vaccines on the list to be available by 2005.

One reason that DoD has taken the lead in the development of these vaccines is that the diseases involved are not common in the developed countries, which severely limits the profitability of the vaccines. This is only partially true in the case of potential drug treatments: antiviral drugs have proven to be highly specific, like vaccines, so development of a drug to treat Ebola virus, for example, is unlikely to be financially attractive to the private sector. Antibiotics, on the other hand, are generally effective against a wide variety of microbial infections, greatly increasing the potential market of any new product. Drug industry spending on developing new antibiotics has been spurred in recent years by the alarming rise in resistance to even the best of current drugs. Information on specific products is tightly controlled for proprietary reasons, but the industry trade group, the Pharmaceutical Research and Manufacturers of America, reports that there are some 125 new antibiotics currently in some stage of development. It seems safe to say that few if any of these are likely to be tested for efficacy against any of the biological agents being considered in this report. Some may nevertheless prove highly effective, and a modest program to screen new antiviral and antimicrobial drugs for activity against biological warfare agents would certainly be a worthwhile R&D investment.

The remainder of this section describes current and potential countermeasures to each of the biological agents on our list and concludes with a description and evaluation of some DARPA-sponsored research into generic, or at least multiagent, countermeasures.

Anthrax

Anthrax is primarily a disease of herbivorous animals, domesticated as well as wild, and humans usually become infected by contact with infected sheep, goats, cattle, pigs, or horses (or contaminated products, for example, wool). The causative agent is *Bacillus anthracis*, a bacterium that forms inert spores when exposed to oxygen. These spores are extremely hardy and may survive outside a living host for years. Infections begin when spores are inhaled, ingested, or enter the body through a skin wound. Germination then occurs and bacteria proliferate. Cutaneous infections produce ulceration at the site, along with fever, malaise, and headache, but mortality is very low with antibiotic treatment. Gastrointestinal infection also begins with fever, malaise, and headache; severe abdominal pain follows, and mortality may be as high as 50 percent. Although military biological weapons programs and speculation about bioterrorism have focused on inhalational infection, naturally occurring cases of inhalation anthrax are rare. In these cases, the initial nonspecific symptoms have been followed by increasingly severe respiratory distress, cyanosis, and shock. Nearly 100 percent of such cases are fatal if left untreated. Meselson et al. (1994) provided extensive documentation of a major outbreak of inhalational anthrax in Sverdlovsk, USSR, in 1979.

Preexposure Prophylaxis

A licensed vaccine with demonstrated efficacy against cutaneous anthrax is available from Michigan Biological Products Institute. This vaccine is the formalin inactivated filtrate from culture of nonencapsulated *B. anthracis*. The principal antigen is Protective antigen (PA), although Lethal factor (LF) and Edema factor (EF) may also be involved in protective immunity. It is administered in six intramuscular doses at 0, 2, and 4 weeks, 6, 12, and 18 months, and affords continued protection if followed by annual boosters. Franz et al. (1997) note that there are few data regarding efficacy against inhalational anthrax in humans, although the vaccine has been shown to provide protection in studies using rhesus monkeys. Although the stockpile is not intended for civilian use, the Department of Defense has approximately seven million doses in cold storage, one million of which are bottled and ready for use (Danley, 1997). Since the release of our interim report, the Secretary of Defense has announced plans to vaccinate all U.S. military personnel. This decision will ensure continued U.S. production capability, but will almost certainly draw down the inventory substantially.

Postexposure Therapy

Penicillin is the recommended treatment of inhalational anthrax, but tetracycline, erythromycin, and chloramphenicol have been used with success (Friedlander, 1997). A variety of other antibiotics have shown *invitro* activity, and current military doctrine calls for initiating treatment with oral ciprofloxacin or doxycycline as soon as exposure to anthrax spores is suspected and introducing intravenous ciprofloxacin at the earliest signs of infection or disease (Franz et al., 1997). It is essential to start antibiotic therapy before or very soon after such signs appear, if a high mortality rate is to be avoided. Other therapies for shock, volume deficit, and adequacy of airway may be necessary. The vaccination series should also be administered to victims not immunized in the previous 6 months.

Antibiotic treatment should be continued for at least 4 weeks (i.e., until at least three doses of vaccine have been received). Penicillin and especially streptomycin are rarely used anymore, and hospital pharmacies will have very limited supplies on hand, but Pfizer will still ship streptomycin overnight. Ciprofloxacin and doxycycline are prescribed far more often, but they are expensive, especially ciprofloxacin, which may limit supplies in any one locale.

Potential Advances

The utility, indeed necessity, of anthrax vaccination *subsequent* to exposure is unique among the biological agents on our list. Anthrax vaccine is thus an exception to the view expressed above that vaccination has limited value as a bioweapon defense for civilians. However, the current vaccine, made by outdated technology, has several disadvantages. It is an impure mixture of bacterial products. Antigen content is variable from lot to lot due to the manufacturing process and the inability to precisely quantify antigenic components. Guinea pig potency assays are only semiquantitative. The requirement for multiple doses is a serious limitation, especially if the vaccine is needed for use in response to exposure of a civilian population.

The current state of knowledge on anthrax pathogenesis and studies of experimental anthrax vaccines indicate that a second-generation vaccine can be developed that could provide protection equal to, or better than, the current vaccine and would require fewer doses. A very effective two-dose vaccine is an achievable goal that should be aggressively pursued through a program that combines research and product development. A single-dose vaccine is a challenging goal that may or may not be achievable.

Research is needed to define the optimal antigenic composition of a new vaccine. A vaccine based on purified protective antigen alone may meet the requirements, but there is a possibility that it will not be the optimal formulation. Including other antigenic components including lethal factor and edema factor and possibly others may enhance efficacy. New adjuvants or new formulations, such as microencapsulation, and alternative delivery systems, such as an oral delivery formulation, should be explored.

Recent publications by Jackson et al. (1998) and Pomerantsev et al. (1997) have raised the question of whether certain strains of anthrax, either deliberately engineered or selected from nature, can overcome the protective immunity generated by a vaccine that is composed principally or entirely of protective antigen. Variation in virulence among anthrax strains and variation in relative resistance to vaccine-induced immunity has been observed in vaccination-challenge experiments in animals, but the basis for the variation is unclear. Antigenic variation in protective factor has been postulated but not demonstrated. The existence of additional virulence factors other than the two plasmid-encoded toxins and the poly-D glutamic acid capsule is a matter of conjecture. The preliminary findings of multiple anthrax strains in the Sverdlovsk anthrax victims by PCR and DNA analysis (Jackson et al., 1998) has raised questions regarding the spectrum of protective immunity provided by current vaccines.

Recently published studies (Pomerantsev et al., 1997) have shown that insertion of the cereolysine AB gene from *B. cereus* into a virulent strain of *B. anthracis* enables the anthrax organism to overcome immunity induced by a live attenuated vaccine strain. Insertion of the cereolysine AB gene into the vaccine strain restores its ability to protect against the modified virulent organism. Questions raised by these studies should be experimentally addressed, within the legal and ethical constraints accepted by the U.S. in this area. Most importantly, will inactivated vaccine containing only protective factor provide protection in man against anthrax strains containing the cereolysine AB gene? Are there additional virulence factors in anthrax strains related to genes homologous with the cereolysine AB gene? Answers to these questions will help guide the design of a second generation vaccine.

Licensure of a second generation vaccine in the absence of any possibility to conduct a formal efficacy trial will require additional studies of the pathogenetic mechanisms and the correlates of protective immunity.

A second-generation, highly effective, and easy to administer anthrax vaccine would substantially improve the nation's ability to protect both civilian and military personnel against the number one biological threat.

R&D Needs

> 8-15 *A vigorous national effort is needed to develop, manufacture, and stockpile an improved anthrax vaccine. This will both benefit the armed forces and enhance the ability to protect the civilian population. The ongoing DoD effort should be supported and accelerated by a well-coordinated complementary DHHS program.*

Brucellosis

Brucellosis is another disease of domesticated animals and usually occurs in humans as a result of ingestion of unpasteurized dairy products. Person-to-person transmission is very rare. The infectious agent is one of six species of the *Brucella* bacterium. Although nonsporulating, brucellae are aerobic organisms viable for long periods outside a host. Its ready transmission by the aerosol route led the United States to experiment with weaponizing *Brucella* during World War II, although the resulting bombs were never used. Fever, chills, and body aches occur in nearly all cases and regardless of route of infection. Brucellae disseminate widely and may cause disease in nearly any organ system, so additional signs and symptoms vary widely. Although rarely fatal, brucellosis can be debilitating for weeks or months if not treated. See Hoover and Friedlander (1997) for additional information.

Preexposure Prophylaxis

There is no approved *Brucella* vaccine for humans.

Postexposure Therapy

According to Franz et al. (1997), patients should be treated with combinations of antibiotics because treatment with a single antibiotic causes poor response or relapse. Usually, a combination of doxycycline and rifampin is given orally for six weeks. Trimethoprim-sulfamethoxazole can be substituted for rifampin, although relapse rates may be as high as 30 percent (Franz et al., 1997). The recommended treatment for bone and joint infections, endocarditis, and central nervous system disease is streptomycin or another aminoglycoside, and therapy should be extended.

Potential Advances

All of the current R&D on brucellosis located by the committee focuses on development of a vaccine. As noted above, the committee considers it unlikely that a vaccine could be usefully employed for protection from a domestic terrorist attack and therefore considers such R&D a low priority for improving civilian medical capability. Antibiotic treatment, though not simple, is possible with current products. USAMRIID conducts assays of second- and third-generation antibiotics as they come on the market, using all of the bacterial threat agents in animal models.

R&D Needs

No action is required at this time.

Pneumonic Plague

Plague is well known as the cause of the Black Death, which devastated the population of Europe in the fourteenth century. The infectious agent is *Yersina pestis*, a nonsporulating bacillus maintained in nature in fleas, most notably the rat flea. In humans the bite of an infected flea leads to a high fever, chills, and headache, often accompanied by nausea and vomiting. Six to eight hours later, very painful swelling of one or more lymph nodes (a *bubo*, hence bubonic plague) develops. Without treatment, septicemia will develop in 2 to 6 days, with a mortality rate of 33 percent. Inhalation of *Y. pestis* aerosol will lead to pneumonic plague (extensive, fulminant pneumonia with bloody sputum), which is almost always fatal if not treated within 24 hours of symptom onset. Patients in terminal stages of pneumonic or septicemic plague may develop large subcutaneous hemorrhages, which may have given rise to the name "Black Death." Additional information is available in McGovern and Friedlander (1997).

Preexposure Prophylaxis

A licensed, killed whole-cell vaccine is available. Although some epidemiologic evidence supports the efficacy of this vaccine against bubonic plague, its efficacy against aerosolized *Y. pestis* has not been established.

Postexposure Therapy

Plague pneumonia is almost always fatal if treatment is not initiated within 24 hours of the onset of symptoms. Streptomycin is administered intramuscularly for 10 days (2 doses each day). Gentamicin can be substituted for streptomycin. Plague meningitis and cases of circulatory compromise are treated with chloramphenicol given intravenously. Intravenous doxycycline administered for 10 to 14 days is also effective.

Potential Advances

With the exception of four projects examining the mechanism of *Y. pestis* virulence factors, the very small amount of current R&D on plague located by the committee focuses on development of a second-generation vaccine. As noted above, the committee considers it unlikely that a vaccine could be usefully employed for protection from a domestic terrorist attack and therefore considers such R&D a low priority for improving civilian medical capability. Antibiotic treatment, though not simple, is possible with current products. USAMRIID conducts assays of 2nd- and 3rd-generation antibiotics as they come on the market, using all of the bacterial threat agents in animal models.

R&D Needs

No action is required at this time.

Tularemia

Tularemia results from infection by the insect-borne bacterium *Francisella tularensis*. In North America, the tick is the principal reservoir, and the rabbit is the vertebrate most closely associated with transmission. As few as 10 organisms can give rise to a clinical infection in humans (Saslaw et al., 1961a, 1961b), and transmission may be via inhalation, ingestion, or, most commonly, through breaks in the skin. The disease is characterized by fever, localized ulceration, enlarged lymph glands, and, in about 50 percent of patients, pneumonia. Without treatment with antibiotics, patients may have a prolonged illness with malaise, weakness, and weight loss persisting for months. Treatment with appropriate antibiotic drugs reduces the duration and severity of the disease, and overall mortality is quite low (1 to 2 percent).

Preexposure Prophylaxis

The United States Army Medical Research and Material Command is the IND holder for a live attenuated tularemia vaccine that appears to be effective against inhalational exposure.

Postexposure Therapy

Streptomycin is administered intramuscularly in two divided doses daily for approximately 10 to 14 days. Gentamicin is also effective. Tetracycline and chloramphenicol are also effective but tend to be associated with significant relapse rates (Franz et al., 1997). See Evans and Friedlander (1997) for additional information.

Potential Advances

The committee could find no active U.S. research on tularemia. Given the possibility of effective treatment with current antibiotics and the recent increase in antibiotic development to counter resistance in many more common pathogens, tularemia research is not a high priority. USAMRIID conducts assays of 2nd and 3rd generation antibiotics as they come on the market, using all of the bacterial threat agents in animal models.

R&D Needs

No action is required at this time.

Q Fever

Q fever is an incapacitating but rarely fatal disease caused by the rickettsia-like agent *Coxiella burnetti*. A large number of mammalian species can serve as host for *C. burnetti*, but humans are apparently the only hosts in which infection results in a disease. Although the organism cannot grow or replicate outside host cells, inhalation of a single organism can result in disease. The usual route of human infection is through contact with domestic livestock, but this may be very indirect contact, because the agent can assume a spore-like form that is extremely resistant to heat, desiccation, and many standard antiseptic treatments, allowing the organism to survive on inanimate surfaces for weeks or months. Human infection is usually the result of inhalation of infected aerosols, and signs and symptoms appear 10 to 40 days after exposure, sometimes abruptly and sometimes very gradually. There is no characteristic set of signs and symptoms, although fever and chills are nearly universal. Headache, fatigue, muscle aches, anorexia, and weight loss are common. Fatalities from Q fever are very rare, and although malaise and fatiguability may persist for months, most other effects last only 2 to 3 weeks. For additional information see Byrne (1997).

Preexposure Prophylaxis

Q fever vaccines in the United States are still investigational, although an effective vaccine, Q-Vax, is licensed in Australia.

Postexposure Therapy

The most common treatments for Q fever are tetracyclines. Macrolide antibiotics, such as erythromycin and azithromycin, are also effective. Other agents used to treat Q fever include quinolones, chloramphenicol, and trimethoprim-sulfamethoxazole. Clinical experience with these drugs is limited. Treatment is most effective when administered during the 10- to 40-day incubation period.

Potential Advances

The committee could locate only two current U.S. studies of *C. burnetti*. Both of the NIH-funded grants are exploring genes and gene products thought to be involved in pathogenesis. Given the possibility of effective treatment with a wide selection of current antibiotics and the recent increase in antibiotic development to counter resistance in many more common pathogens, Q-fever research is not a high priority. USAMRIID conducts assays of 2nd- and 3rd-generation antibiotics as they come on the market, using all of the nonviral threat agents in animal models.

R&D Needs

No action is required at this time.

Smallpox

Until very recently, smallpox was an important cause of morbidity and mortality in the developing world. The causative agent of smallpox is variola, one of a family of large, enveloped deoxyribonucleic acid (DNA) poxviruses. Unlike many of the agents discussed above, the variola virus thrives only in human hosts, and as a result, aggressive case finding and vaccination programs (using the closely related but nonpathogenic vaccinia virus) are thought to have eradicated smallpox. The last known cases occurred in 1978. Concerns about its use as a weapon persist, however, because variola virus is highly stable and retains its infectivity for long periods outside the host, and because enough is known of its sequencing that biotechnology might be used to create variola or a pathogenic variation of variola. Although characteristic pustular skin lesions provided the name for this disease, and virus can be recovered from scabs throughout convalescence, smallpox is infectious by aerosol as well. It is transmitted more easily than any of the other agents being considered in this report, and its use in a terrorist attack would pose the threat of a global epidemic. Regardless of route of transmission, clinical manifestations begin with fever, malaise, headache, and vomiting, and the infection is a systemic one that produced mortality rates of 20 to 30 percent in unvaccinated populations (McClain, 1997).

Preexposure Prophylaxis

Individuals who were vaccinated during the WHO smallpox eradication campaign in the 1970s were considered to have immunity to smallpox for at least 3 years, but protection diminishes over time. The only vaccine still available in the United States is a live vaccinia virus manufactured by Wyeth-Ayerst Laboratories (now Wyeth-Lederle Vaccines and Pediatrics, and no longer manufacturing the vaccinia vaccine). The CDC holds the entire remaining stock (approximately 6 million doses). Vaccination causes a pustule and local reaction on intradermal administration. Adverse reactions include encephalitis and other neurological disorders, generalized vaccinia, and vaccinia necrosum. Virus is transmissible to nonvaccinees and hazardous to individuals with eczema or immunosuppressive disorders.

Postexposure Therapy

Vaccination will give protection to an exposed individual if it is administered within a few days of exposure, regardless of time since any prior vaccination. There is no chemotherapeutic agent with proven effectiveness against smallpox, but Franz et al. (1997) suggest that preclinical tests against other poxviruses indicate that chemotherapy with cidofovir might be useful (see below). Vaccinia-immune globulin (VIG) may also be of use if given within the first week following exposure (preferably within 24 hours). VIG, which is prepared from the blood of repeatedly vaccinated persons, is available from the CDC in extremely minute quantities. Because almost no one is being vaccinated anymore, there is little prospect of producing a large stockpile of VIG.

Potential Advances

Smallpox presents a unique risk among the possible biologic weapons in that the secondary contamination risk (person-to-person transmission) is significant. This is in distinction of virtually all other candidate biologic weapons. Coping with a global pandemic produced by use of this weapon would require significant investment in research and development of vaccine and antiviral therapies as well as a significant investment of public health resources. A careful risk assessment as to the likelihood of this agent being employed would guide the appropriate response required.

Vaccination is the only proven and feasible means of combating an epidemic that could result from deliberate release of smallpox virus in the U.S. population. Antiviral drugs may be of value in dealing with infected patients, but expanding ring vaccination is the proven means of eradicating foci of infection. Smallpox vaccine is therefore the second (and last) exception to the committee's strong preference for treatments in planning for terrorist incidents rather than preexposure prophylaxis. The current U.S. stockpile of vaccine is far less than would be needed in the event of such a contingency. There is no existing licensed manufacturing capacity for production of additional stocks of the current vaccine. Use of the current vaccine would entail a substantial risk of vaccine-induced complications, many of which would require treatment by vaccinia-immune globulin and antiviral drugs. The epidemic of AIDS substantially increases the risk, since generalized vaccinia is known to occur in vaccinated AIDS patients. Reestablishing manufacturing of the current product is not a recommended option in view of the undesirable characteristics of the product and the potential for improvement. An experimental vaccine has been under development by the U.S. Army and is included in the current DoD Joint Vaccine Acquisition Program. This candidate vaccine contains a virus derived from a previously licensed vaccinia strain (Connaught strain), has been produced in cell culture, and has progressed to phase one trials. Licensure of a new vaccine in the absence of any possibility to conduct efficacy trials will pose multiple research problems, especially those relating to correlates of vaccine-induced immunity and levels of protection. Although unlikely to solve all of the problems of traditional vaccinia vaccine, development and manufacture of adequate stockpiles of this vaccine is the current best option for dealing with the contingency of a release of smallpox virus.

R&D Needs

> *8-16 The development, manufacture, and stockpiling of an improved smallpox vaccine for post-attack management of a potential epidemic should be given a high priority. DHHS agencies could assist the military development program by addressing research questions related to product development, such as correlates of immunity. An agreement with the DoD and the manufacturer of a new vaccine on purchase and stockpile of vaccine for civilian use may be an important incentive and an important factor in production planning.*

Recent research by scientists at NIH and USAMRIID on antiviral drugs against orthopox viruses, including variola, have shown some promising leads, including antivariola activity by at least three classes of compounds (Huggins et al., 1996; 1998). One of these includes a licensed drug, cidofovir (marketed as Vistide™), currently used to treat cytomegalovirus retinitis in AIDS patients. However, cidofovir is an intravenous preparation with substantial toxicity and would therefore be of limited value in the event of a terrorist release of variola virus. Retention of U.S. stocks of variola virus currently scheduled for destruction in 1999 would be of value to a drug discovery and development program. Pox viruses vary widely in their sensitivity to chemotherapeutic agents, and use of surrogates for variola such as monkeypox cannot be relied upon totally, although monkeypox is a serious emerging disease in central Africa, and development of a drug that is active against variola and monkeypox would provide an immediate benefit independent of terrorism. Similarly, development of an antipox drug that could treat vaccinia problems, such as eczema vaccinatum and generalized vaccinia in the immunosuppressed, could be of great value in the event we need to do large-scale vaccination in the future. Notwithstanding the possible additional benefits of focusing drug development on variola surrogates, direct *in vitro* testing against variola virus is important to be sure of usefulness in treating smallpox.

R&D Needs

> 8-17 *A major R&D program should be undertaken to exploit the previous studies to discover and develop new antismallpox drugs for therapy and/or prophylaxis.*

Viral Encephalitides

Although other viruses can also produce encephalitis, three closely related enveloped RNA viruses of the Alphavirus genus initially recovered from moribund horses in the 1930s are considered the primary candidates for weaponization: Venezuelan equine encephalomyelitis virus (VEE), eastern equine encephalomyelitis virus (EEE), and western equine encephalomyelitis (WEE). All could be inexpensively produced in quantity, are relatively stable, and are readily amenable to genetic manipulations that might confound defenses against them. Natural infections are acquired through mosquito bites, but these viruses are also highly infectious as aerosols. Victims develop an incapacitating combination of fever, headache, and fatigue, and the most severe of the three, EEE, results in case fatality rates of 50 to 75 percent. Survivors may be left with seizures, sensorimotor deficits, or cognitive impairment. See Smith et al. (1997) for additional information.

Preexposure Prophylaxis

A live attenuated vaccine for VEE (TC-83) is immunogenic in 80 percent of recipients, but it causes more than 20 percent of recipients to experience high fever, malaise, and headache serious enough to require bed rest. Inactivated vaccines for VEE, WEE, and EEE in humans also exist; they require multiple injections and have poor immunogenicity. All vaccines, including TC-83, are available only in IND status.

Postexposure Therapy

No specific therapy exists for these alphavirus encephalitides and treatment is directed at management of specific symptoms (e.g., convulsions, respiratory infection, and high fever). Even treatment with virus-neutralizing antisera (antibody-containing serum from the blood of previously-infected patients or animals) will fail to stop progression of established encephalitis. Antimosquito precautions should also be implemented.

Potential Advances

As is the case with most of the other infectious diseases of concern to the biological defense program, the primary thrust of current R&D is on vaccine development, and several candidate vaccines using live attenuated VEE, WEE, and EEE viruses have been identified and tested in animal models at USAMRIID. Unlike the antibiotics used to treat bacterial infections, most antiviral drugs are highly virus-specific, so drugs like AZT and the protease inhibitors that have proven so successful in controlling HIV, for example, have not been useful against the viral encephalitides. For the same reason, there is little incentive for drug companies to pursue an antiviral drug for VEE, WEE, or EEE, diseases for which there is essentially no market in the developed nations. The potential market for a broadly effective antiviral drug is huge, however, and DARPA is sponsoring a number of research projects aimed at structures or processes common to a number of different viruses. For example, scientists of enVision and Boston Biomedical Research Institute have isolated developmental proteins which regulate cell proliferation in animal fetal tissues (Barnea et al, 1995; 1996) and have begun testing them for activity against viruses (whose replication is inherently linked to the host cell).

A project at the University of Wisconsin focuses on design of compounds that inhibit viral entry, intracellular transport, maturation, and release. Combinatorial chemistry and organic synthesis will be used to design compounds to prevent progression of infection by viruses with bioterrorism potential.

Teams at the University of Texas Medical Branch and the University of Wisconsin are working jointly using combinatorial chemistry to design antiviral drugs that will act by inhibition of capsid-RNA interaction, polymerase activity, or glycoprotein attachment to cellular receptors. A group at the University of Alabama, Birmingham is also using combinatorial chemistry to design a drug to inhibit capsid-RNA interaction (Edberg and Luo, 1997; DeLucas, 1998), and scientists at The Scripps Research Institute are attempting to build antiviral antibodies that will enter infected cells and fight viruses at the intracellular level (McLane et al., 1995; LeBlanc et al., 1998).

Finally, GeneLabs Technology is attempting to develop broad-spectrum antiviral drugs from a large library of chemical compounds by assaying for RNA binding and selecting for dimer molecules that bind dsRNA, but not DNA.

All of these projects, funded as they are by DoD, begin with the aim of improving biological defense, so, although most are years away from a licensed product, effectiveness against viruses like VEE will be a central feature rather than an adventitious side effect.

R&D Needs

> 8-18 *Support for broad-spectrum antiviral drugs for treatment of VEE, WEE, EEE and other viruses considered biological terrorism threats should be considered a high priority.*

Viral Hemorrhagic Fevers

Viral hemorrhagic fever is a term indicating an acute febrile illness accompanied by circulatory abnormalities and increased vascular permeability. Similar diseases result from infection with any of about a dozen RNA viruses belonging to four different families: *Arenaviridea* (Lassa, Argentine, Bolivian, Venezuelan, Brazilian), *Bunyaviridea* (Rift Valley, Crimean-Congo, and Hantaan), *Filoviridea* (Marburg and Ebola), and *Flaviviridea* (Dengue and Yellow Fever). All of these diseases are thought to be transmitted to humans through contact with infected animal reservoirs or arthropod vectors (mosquito or tick). All are relatively stable and highly infectious as fine-particle aerosols. Patients generally present with high fever and some indication of vascular involvement: low blood pressure, flushing, or small subcutaneous hemorrhage. Progression of the disease typically involves bleeding from mucous membranes, signs of pulmonary, liver, or kidney failure, and shock. Mortality varies widely among the diseases, from 5 to 20 percent of symptomatic cases for most, but as high as 90 percent for Ebola virus. See Jahrling (1997) for further information.

Preexposure Prophylaxis

Vaccines are available for yellow fever (YF), Rift Valley fever (RVF), and Argentine hemorrhagic fever (Junin virus, JUN). Cross protection against Bolivian hemorrhagic fever may also be provided by the Junin vaccine. These are the only vaccines available for any of this set of diseases and will be effective only if personnel are immunized before exposure. Only yellow fever vaccine is licensed by the FDA; the others are used in the United States under IND protocols and can only be obtained through the CDC.

Postexposure Therapy

Vaccines have no application in treatment of exposed targets. Intravenous administration of the antiviral drug ribavirin is recommended for therapy of infections with Lassa virus and with Hantaan and other Old World Hantaan-related viruses. Ribavirin may also be useful for treatment of infections by other arenaviruses and Crimean-Congo hemorrhagic fever (CCHF) virus, but data proving efficacy are lacking. Ribavirin for these infections is used under IND protocols. It is not thought likely to be effective against filoviruses, such as Ebola, or flaviviruses, such as YF or Dengue. There is no proven chemotherapeutic drug available. Human immune serum is efficacious for treatment of persons exposed to Junin virus. Some anecdotal evidence suggests that Ebola human convalescent serum may be effective in preventing death from Ebola virus, but no scientifically controlled studies have been reported. Case management includes careful monitoring of fluid and electrolytes and intravenous corrective therapy where needed.

Hospitalization under barrier precautions (gloves and gowns, face shields, or surgical masks and eye protection, for all those coming within 3 feet of the patient) is usually adequate to prevent transmission of Ebola, Lassa, CCHF, and other hemorrhagic fevers, but isolation of the patient provides an added measure of safety and is preferred, if facilities are available. Disinfection of bedding, utensils, and excreta by heat or chemicals is recommended for all of the viral diseases under consideration. Quarantine, defined by Benenson (1995) as "restriction of the activities of well persons or animals who have been exposed to a case of communicable disease during its period of communicability, to prevent disease transmission during the incubation period if infection should occur," may be indicated following an act of bioterrorism. If the agent is already identified, the decision to quarantine should be made based on the known communicability of the agent. Quarantine, for instance is not recommended for those exposed to anthrax, but is recommended for those exposed to plague, if chemoprophylaxis is not available. If the agent is not identified, then quarantine should be considered. CDC has provided detailed instructions on the management of suspected hemorrhagic cases, including handling and laboratory testing of potentially infectious materials (Centers for Disease Control and Prevention, 1988, 1995b).

Potential Advances

CDC, USAMRIID, and NIH all support small programs of research on one or more of the hemorrhagic fever viruses, primarily basic research on mechanisms of pathogenicity and explorations of possible vaccine candidates. The collaboration between USAMRIID and the NIAID Drug Discovery Program that identified cidofovir as a potential smallpox treatment has also discovered a class of compounds (s-adenosyl homocysteine hydrolase inhibitors) that may be effective against filoviruses, such as Ebola. In a mouse Ebola model that produces 100 percent mortality within 7 days, treatment beginning on the day of exposure provided 100 percent protection, and treatment beginning 4 days after exposure saved 40 percent of infected mice (memo from John Huggins of USAMRIID to F Manning, May 21, 1998).

Much of the discussion of antiviral drug therapy for the viral encephalitides is also applicable in the case of the hemorrhagic fever viruses. The apparent differences in the activity of ribavirin against the various viruses of this group further underlines the specificity of current antivirals and emphasizes the need for broad-spectrum compounds. In addition to the DARPA-sponsored work cited in that discussion, researchers at Inotek have developed a compound that inhibits the nitric oxide pathway at multiple sites. Although this drug has potential for broad application to infectious agents that cause oxidative damage as part of their pathogenesis, it is being tested for efficacy in the Ebola guinea pig model. Additionally, arenavirus anti-polymerase humanized monoclonal antibodies will be synthesized at the University of Wisconsin for evaluation as antiviral drugs and researchers at the University of Texas Medical Branch are attempting to find ways to inhibit the intracellular transcription factor NFkB to modulate cytokine effects that are associated with arenavirus pathogenicity.

R&D Needs

8-19 Support for the discovery and development of antiviral drugs for treatment of viral hemorrhagic fevers and other viral diseases considered biological terrorism threats should be considered a high priority.

Botulinum Toxins

Botulism, an often lethal form of poisoning associated with improperly canned or stored foods, is the result of neurotoxins produced by the spore-forming anaerobic bacterium *Clostridium botulinum*. The botulinum toxins are the most toxic substances known. Some have an estimated LD_{50} of 1 nanogram/kilogram of body weight (Gill, 1982). Some *in vitro* work suggests that these neurotoxins act presynaptically to block the release of acetylcholine and perhaps other neurotransmitters (Habermann, 1989), but the exact mechanism is as yet unknown. It is known that whether ingested, inhaled, or injected, the clinical course is similar. Several hours to 1 or 2 days later, dry mouth, difficulty swallowing, and double vision may be reported, followed by a progressive muscle weakness culminating in respiratory failure from skeletal muscle paralysis. See Middlebrook and Franz (1997) for additional information.

Preexposure Prophylaxis

The currently available vaccine is a formalin-fixed supernatant from cultures of *C. botulinum*. It protects against botulinum toxin types A through E, but is available only as an IND product, with a license held by the CDC. A series of three vaccinations must be started 12 weeks before exposure, and 80 percent of recipients exhibit protective titers at 14 weeks. Yearly boosters are required to maintain protection. Although it is an IND product, the vaccine has been given to hundreds of people since its development in the 1950s.

Postexposure Therapy

Foodborne botulism is treated with a licensed trivalent equine antitoxin (serotypes A, B, and E) that is available only from the CDC. There is no other approved therapy for airborne botulism, although animal studies show that botulinum antitoxin can be very effective if given before the manifestation of clinical signs of disease. Mechanical ventilation is invariably necessary due to paralysis of respiratory muscles, if antitoxin is not given before the onset of clinical signs (Shapiro et al., 1997).

Potential Advances

See Table 8-6. A despeciated equine heptavalent antitoxin that has been developed by the U.S. Army specifically for aerosol exposures to serotypes A–G has IND status (Franz et al., 1997). This antitoxin markedly reduces the chances of serum sickness by eliminating the species-specific antigens from the horse immunoglobin (the basic immunoglobulin molecule is altered by removing complement fixing (Fc) region with pepsin to produce a fragment labeled $F(ab^1)_2$). This $F(ab^1)_2$ antitoxin protected animals from inhaled toxin at doses of 10 times the LD_{50} when given prior to exposure and was also fully protective when given after exposure, as long as it was given prior to the onset of clinical signs. Further DoD work focuses on developing a recombinant vaccine, with much less substantial investigations of monoclonal antibodies and drugs to inhibit toxin uptake into cells (metalloprotease inhibitors).

TABLE 8-6 Potential Countermeasures for Botulism

Antidote	Efficacy	Availability	Potential Civilian Utility	Stockpile
Active immunization: vaccine (toxoid of serotypes A–E) Formalin-fixed crude culture supernatant from strains producing appropriate serotypes vaccination 0, 2, and 12 weeks with annual booster.	~ 80% of recipients exhibit protective titers (CDC standard > 0.25 IU/ml) at 14 weeks; at 1 year booster almost none with measurable titer. Booster results in 90% of individuals with good response. Fully protects animals from all routes of challenge.	Investigational (IND)	Preexposure Task force	Health dept.
Passive immunization: Horse antibotulism serum (globulin) A despeciated globulin treated with Pepsin to produce $F(ab^1)_2$ Basic immunoglobulin molecule altered by removing complement-fixing (Fc) region to concentrate antigen binding sites.	Decreases risk of serum sickness of older equine version: Studied in monkeys with serotype A prevents disease development if pretreated. When given after exposure protective if given before the onset of clinical signs. If given after the onset of symptoms, no protection.	Research	Yes Postevent Prehospital and postexposure high-risk	Health dept.

Recombinant vaccines	Will enhance safety	Investigational		N/A
Monoclonal antibodies	Will enhance safety	Investigational		N/A
Aminopyridines (3,4-diaminopyridine)[a]	Limited success with prevention and reversal of muscle paralysis for human serotype A. in animal studies. No effect for other serotypes. High toxicity potential.	Not under further study.		N/A
Chimer of receptor binding protein for botulinum molecule (either monoclonal antibody or drug to neutralize intracellularly)	Preclinical studies only	Research		N/A
Botulism immune globulin harvested from human donors experimentally exposed to toxoid[b]	Longer biological half life with prolonged effective level	Investigational	Preexposure, postevent Prehospital and postexposure high-risk	Health dept.

[a]Siegel et al., 1986; [b]Frankovich and Arnon, 1991.

R&D Needs

> *8-20 Recombinant vaccines, monoclonal antibodies, and antibody fragments all have potential benefit, but require extensive investigation. Advances in these areas would benefit unintentional botulism case care, and the effort could provide prophylaxis and early treatment for those exposed to a potential toxin in a foodborne epidemic. Investigation of new techniques for this disorder would thus have substantial societal benefit for a rare clinical occurrence that is also a possible biological warfare event.*

> *8-21 Further investigation into the utility of botulinum immune globulin would offer an immediate post-exposure therapy. This would be a great advance, because current passive immunity is equine derived and probably only beneficial prior to the onset of symptoms. Preexposure active immunization might be beneficial for those with the potential to be placed at very high risk, such as Hazmat or MMST teams, but this appears to be a very low-probability occurrence.*

Staphylococcal Enterotoxin B (SEB)

SEB is one of seven toxins produced by strains of the *Staphylococcus aureus* bacterium. Like botulinum toxin, it is most often associated with food poisoning. Unlike botulism neurotoxin, SEB appears to exert its effects through overstimulation of cytokine production by the immune system (Ulrich et al., 1997). Ingested SEB is incapacitating rather than lethal, with vomiting and nausea prominent. It is relatively stable in aerosol, however, and the consequences of inhalation may be much more severe, possibly even a fatal "toxic shock" syndrome involving high fever, a rapid drop in blood pressure, and multiple organ failure.

Preexposure Prophylaxis

No human vaccine against SEB is available, although there are several vaccines in development (Ulrich et al., 1997).

Postexposure Therapy

Because there is no approved antitoxin, therapy is currently limited to supportive care focused on reductions of fever, vomiting, and coughing. Respiratory symptoms may follow exposure to aerosolized SEB, and mechanical ventilation may be necessary. Most patients usually recover in 1 to 2 weeks without residual effects.

Potential Advances

Efforts to better characterize the mechanism of action of SEB and related "super antigens" are being funded by NIH (at a very modest level), and USAMRIID is pursuing both passive and active immuno-protection. Administration of chicken-derived anti-SEB antibodies prior to, or up to 4 hours after, inhalation of an otherwise lethal dose of SEB protected nonhuman primates from death (but not from illness). Active immunization is felt to be the most promising line of defense due to the very rapid binding of the toxin (less than 5 minutes).

R&D Needs

8-22 Given the questionable utility of active immunization for civilian first responders and the generally nonlethal effects of the toxin, the committee recommends that SEB research continue to be considered a relatively low priority.

Ricin

Ricin is a protein found in the bean of the castor plant, *Ricinis communis*, which has been widely cultivated for its oil since ancient times. Ricin remains in the castor meal after the oil is extracted, but is readily separated and concentrated. Although its lethal toxicity is about 1,000-fold less than that of the botulinum toxins, the worldwide availability of large quantities of castor beans makes ricin a potential biological weapon. At the cellular level, ricin kills through inhibition of protein synthesis. Clinical signs and symptoms appear 8 to 24 hours after exposure and vary with the route of exposure: respiratory distress and airway lesions after inhalation; vomiting, diarrhea, gastrointestinal, liver, and kidney necrosis after ingestion. Ricin is not dermally active. Although intramuscular injection of ricin was used in the highly publicized assassination of Bulgarian defector Georgi Markov in 1978 (Crompton and Gall, 1980), little human data exist on mortality rates after ricin poisoning by the aerosol route. The death rate in cases of castor bean ingestion has been low, under 10 percent (Rauber and Heard, 1985).

Preexposure Prophylaxis

Although preclinical testing in animals has encouraged the U.S. Army to submit an IND application to the FDA for a formalin-treated toxoid immunization, no human testing has been conducted, and no vaccine is available for clinical use.

Postexposure Therapy

Activated charcoal lavage may be helpful immediately after ingestion of castor beans or ricin, but ricin acts rapidly and irreversibly, which makes treatment very difficult after signs and symptoms appear. Symptomatic care is the only intervention presently available to clinicians treating aerosol ricin poisoning. Additional information on ricin may be found in Franz and Jaax (1997).

Potential Advances

There are several investigational antiricin strategies being pursued (see Table 8-7), including passive immunization through antibodies and toxoid-stimulated immunization, although, as with SEB, the very rapid binding of ricin regardless of the route of challenge makes active immunization the preferred strategy. Both toxoids of the native toxin and a preparation of the A-chain fragment have been shown to provide mice with protection from lethal aerosolized doses. DoD is pursuing the testing of the A-chain antigen as a possible vaccine. USAMRIID has been unable to demonstrate effective passive (postexposure) immunization however, and an *in vitro* screening program has examined over 150 compounds of a wide variety, but has not found a compound that provides any protection to laboratory animals. Additional study of the toxin's mechanism of action may provide useful leads for specific mediator blocking agents, but these research ventures are in a very early stage and raise fundamental questions about risk, benefit, and potential utility, in view of the exceptionally low potential for mass exposure.

R&D Needs

8-23 *Continued investigation of antiricin antibodies as well as formalin-treated toxoid immunization is appropriate, but should be considered low priority for domestic preparedness due to the high cost of developing a licensed product, the limited potential of mass exposure to ricin, and the low probability of any potential means of developing a mass-exposure technique.*

T-2 Mycotoxin

Mycotoxins are by-products of fungal metabolism. A wide variety of fungi produce substances that produce adverse health effects in animals and humans, but mycotoxin production is most commonly associated with the terrestrial filamentous fungi called molds. T-2 mycotoxin is one of a family of nearly 150 toxins produced by *Fusarium* and related fungi

TABLE 8-7 Potential Antidotes for Ricin Poisoning

Antidote	Efficacy	Availability	Potential Civilian Utility	Stockpile
Antiricin rabbit antibodies tested parenterally in mice following inhalation at 1 hour Produced by toxoid of native toxin or purified A chain	Pretreatment at 1 hour protected mice from $5 \times LD_{50}$	Preclinical testing	Prehospital task force	Poison center Health dept.
Toxoid is micro-encapsulated in galactide-glycolyde microparticles	Active immunization	Preclinical testing	Prehospital task force	N/A
Formalin treated toxoid	Passive prophylaxis	Preclinical testing	Postexposure	N/A
α-deglycosylated A chain as antigen	Effective in mice after one immunization	Submitted to FDA as IND	Pretreatment	N/A
Supportive chemotherapy	Transition state inhibitors block enzymatic effects on A chain	Preclinical testing	Insufficient evidence	N/A

that infect wheat and other grains that are important human foods (T-2 mycotoxin-contaminated grain is thought to have been responsible for the deaths of more than 10 percent of the population of the Russian town of Orenburg in the 1940s). These toxins are nonvolatile, low-molecular-weight (250–550) compounds that are insoluble in water and highly resistant to heat. T-2 toxin has been the most extensively studied. Its primary toxic effects appear to be caused by inhibition of protein synthesis. Clinical effects of acute exposure, in addition to local effects specific to route of exposure (unlike the other biological agents described here, T-2 mycotoxin can penetrate intact skin [Wannamacher et al., 1991; Wannamacher and Wiener, 1997]), include vomiting and diarrhea, weakness, dizziness, ataxia, and acute vascular effects leading to hypotension and shock. In the 1970s, the United States government accused the Soviet Union and its allies of using trichothecene mycotoxins as weapons in conflicts in Southeast Asia and Afghanistan (Ember, 1984). See Wannamacher and Wiener (1997) for additional details.

Preexposure Prophylaxis

No vaccine is currently available for protection against any of the trichothecene mycotoxins.

Postexposure Therapy

No specific therapy for trichothecene mycotoxin poisoning is currently available. Skin decontamination with soap and water or the hypochlorite- (M258A1) or resin-based (M291) military decontamination kits can effectively remove toxin up to six hours after exposure, although none of them neutralize the toxin. Treatment of respiratory, dermal, and GI effects currently must be symptom based and supportive in nature. Superactive activated charcoal, for example, a common treatment for many orally taken poisons, has been shown to bind 0.48 mg T-2/gm charcoal in mice and improve survival rates significantly.

Potential Advances

Two topical skin protectants in development by DoD have been shown to protect rabbits from the dermal effects of T-2 toxin for at least 2 hours, but neither is available for human use. An IND has been submitted to the FDA for the simpler of the two, which offers only passive protection from T-2 toxin and a number of other potential chemical and biological agents. Human safety has also been demonstrated for this product, a manufacturing contract awarded, and an NDA is being prepared in hopes

TABLE 8-8 Potential Countermeasures Against T-2 Mycotoxin

Antidote	Efficacy	Availability	Potential Civilian Utility	Stockpile
Topical skin Protectant (TSP)	Passive protection	Goal is FDA license by FY00	Prehospital high-risk personnel	Health dept.
Reactive TSP (decontaminates)	Proof of principle	Goal is FDA license by FY08	Prehospital high-risk personnel	N/A
Corticosteroids (systemic)	High doses decreased primary injury and shock in animal studies	Yes	Possible supportive therapy	N/A
BN5021 (a platelet activating factor antagonist)	Prolongs rat survival when given after a lethal dose	Research	Possible therapy	N/A
Despeciated monoclonal Anti-idiotype antibody	100% effective for rats 30 min before T_2 exposure or 15 min after exposure	Preclinical	Prehospital high-risk personnel	N/A
Prophylactic enzyme induction: flavonoids, ascorbic acid, vitamin E, selenium	Rodent studies only	Preclinical	Prehospital high-risk personnel	N/A
Prophylaxis: mycotoxins conjugated to a carrier protein creates immunogen	Highly specific and not successful against other trichothecenes	Preclinical	Prehospital high-risk personnel	N/A

of fielding the product in the next two years. However, it is difficult to envision a domestic civilian terrorism incident in which these skin protectants would play an important role. Unless police, fire, and rescue personnel were to use a protective lotion routinely, some preincident intelligence would be required, and that should rightly trigger the use of protective clothing and respiratory equipment. Other prophylactic measures that have some promise in preclinical studies are susceptible to the same criticism. Very few specific postexposure measures have received extensive study. To the committee's knowledge, none are currently being studied.

R&D Needs

> *8-24 The committee considers the threat of a terrorist incident involving T-2 mycotoxin to be very low. In addition, its effects are not consistently fatal, nor are they so rapid that prehospital treatment is demanded. The committee therefore recommends that civilian medical personnel continue to rely on nonspecific treatment and supportive therapy. R&D in this area should be limited to screening antivesicant treatments for their efficacy in animal models of mycotoxin poisoning.*

Broad Spectrum Defenses Against Biological Agents

As noted above in the sections on viral encephalitides and viral hemorrhagic fevers, research on a number of multiagent defense approaches is being sponsored by the Defense Advanced Research Projects Agency (DARPA) contract program on Unconventional Pathogen Countermeasures. Exploratory work is under way not only on new antiviral drugs but also on antibacterials, antitoxins, new types of immunization, and several multipurpose approaches to pathogen destruction.

The first of the multipurpose approaches is a strategy that utilized the red blood cell membrane outer surface as a platform for enzymatic-based defenses against pathogens and toxins. Boston University researchers are mounting a toxin-specific enzyme on the surface of circulating red blood cells in order to encounter the toxin and destroy it before the infecting agent can reach the target cells. A similar tack is being taken by University of Virginia scientists, who are using pathogen-specific cross-linked, bispecific monoclonal antibody complexes bound to a red blood cell complement receptor (Taylor et al., 1997; Nardin et al., 1998). The modified circulating red blood cells then scavenge for the infectious agent and bind it. The pathogens and the complement receptor, but not the red blood cells to which they are bound, are rapidly cleared from the circulation and destroyed by hepatic macrophages. The third multi-purpose approach

(Osiris Therapeutics, Inc.) involves engineered human mesenchymal stem cells programmed to produce a specific signal in the presence of an infecting agent or a toxin. The resulting signal would then tell a secondary cell system (i.e., platelets) to release a genetically programmed detoxifying substance. The fourth multi-purpose approach (Genelabs) focuses on the development of self-assembling RNA and DNA binding agents with broad spectrum anti-pathogen effects

These approaches are multipurpose in the sense that each can be tailored to any one of many potential agents, although the agent must be specified. Therefore, the approaches could be used prophylactically if there is prior knowledge of the infecting agent and could be used as treatment after the agent is detected and identified.

Immunizations

The human host's immunity can stop or slow the spread of infection if the immune system can be activated very rapidly after the exposure. Approaches include stem cells (Osiris) for delivery of several antigens, simultaneously or over time, to immunize against a broad range of potential agents and the use of peptides (University of Connecticut) attached to heat shock proteins that together efficiently elicit cytotoxic T-lymphocytes (killer white blood cells) against specific infectious agents (Blachere et al., 1997). Heat shock proteins are produced naturally when the body is stressed by heat or injury; they help cells repair damage from stress. The heat shock protein complexes can be extracted as noninfectious entities from infected cells in culture and used as a vaccine. One does not need to know the identity of the agent, because peptides of any pathogen will putatively work.

Conventional immunization methods take days to weeks to induce protection. There is a race between the body's immune system and the infecting agent. If the development of immunity could be hastened, the body would have a better chance to win. In an attempt to do just that, the University of Texas, Southwestern proposes to remove antigen-presenting cells from the blood, inject them by a special gun with DNA encoding the pathogen's genes, then return them to the body to provide rapid protection (Barry and Johnston, 1997; Johnston and Barry, 1997). Another effort is a three-pronged attack on viruses at Massachusetts General Hospital, in which soluble factors from human immune cells are to be developed to induce antiviral defense; immune effector cells engineered to recognize and destroy virus-infected cells, and hematopoietic stem cells (immature cells from the bone marrow) engineered to give long-term pathogen immunity (Scadden, 1997; Gardner et al., 1997).

Antibacterial Drugs

The development of new antibacterial compounds is of prime importance, especially for prevention of infection when a terrorist attack is suspected, and for treatment after an attack. New classes of drugs are especially needed to combat known bacteria that have been rendered resistant to currently available antibiotics by genetic engineering. The DARPA program is supporting several efforts, including combinatorial chemical technology at ISIS Pharmaceuticals to find decoys that will prevent bacterial RNA-protein interactions (Konings et al., 1997); a Stanford effort to develop decoys that will target a DNA methylating enzyme (McLane et al., 1995; LeBlanc et al., 1998); and gene products expressed early in infection that can be inhibited by decoy compounds (SmithKline Beecham).

A new class of drugs is under development at Stanford University to render pathogenic bacteria nonpathogenic by inhibiting the type III secretory pathway (Mecsas et al., 1998). The drugs will not necessarily kill the bacteria, but will stop the transport of virulence-causing proteins destined to be secreted through the bacterial surface.

Another approach (Harvard University) aims at blocking attachment of toxins, viruses, and bacteria to target cell surfaces. Organic synthesis of strings of polyvalent molecules that mimic specific receptors and that nonspecifically rely on hydrophobic or coulombic interactions is expected to yield compounds that can be administered to persons during a bioterrorist attack.

Antiviral Compounds

New approaches to broad-spectrum antiviral drugs were discussed above in the sections on viral agents and will not be reviewed again here, other than to note that a variety of approaches aimed at common elements of viral replication and pathogenesis are targeted.

Antitoxins

New antitoxins are proposed for development by researchers at Los Alamos National Laboratory, using structure-based design of compounds that block SEB-host interactions at the receptor site on the target cell. A second approach is designed to select low molecular weight peptides from random libraries and test by *in vitro* assays to detect inhibition of cholera toxin binding. Development of broad spectrum molecular antagonists and vaccines is also proposed by Hebrew University scientists working with SEB and other superantigen toxins. These antitoxins may be useful to first responders or emergency departments in cases of toxin exposure, whatever the source. Finally, a Rockefeller University project seeks to express detoxifying enzymes and neutralizing antibodies on the surface or secreted from human commensal bacteria (e.g., *S. gordonii*).

Feasibility and Utility of the DARPA Approaches

DARPA projects are as a rule speculative, but with the potential for high yield when they succeed. All of these approaches are in the initial experimental stage, making it unlikely that the research will mature to an application for several years. Some of the approaches, especially those using combinatorial chemistry (e.g., SELEX, Systematic Evolution of Ligands by Exponential enrichment) for drug development, have already shown great promise in the development of HIV/AIDS drugs (e.g., Tuerk and MacDougal-Waugh, 1993). However, getting similar drugs for bioterrorist organisms to licensure is problematic. The FDA process as it is now structured will need to accept surrogate outcome measures for both bacterial and viral infections, because these infections are for the most part exotic and do not occur commonly in the United States. Furthermore, for the same reason, pharmaceutical manufacturers are not likely to invest heavily in drugs for which the demand is small or infrequent.

R&D Needs

> 8-25 *Development of new specific and broad-spectrum antibacterial and antiviral compounds should be encouraged through financial support, early and accelerated transition to the marketplace, and, where indicated, application of orphan drug coverage. Emphasis should be given to rational drug development through combinatorial chemistry and applied research on receptors, replication complexes, and host defense mechanisms, keeping in mind that for combating terrorist attacks, treatment rather than prevention will be the most practical approach.*

SUMMARY OF R&D NEEDS

Unlike the other chapters of this report, the committee's recommended R&D needs for drugs and other therapies have been agent-specific. This was necessary because the nature of research tends to be agent specific. Additionally, because research of each agent is at various stages, the committee has prioritized the R&D needs as listed below.

High Priority

Nerve Agent

- Antidote stockpiling and distribution system
- Scavenger molecules for pretreatments and immediate postexposure therapies

Vesicants

- An aggressive screening program focused on repairing or limiting injuries, especially airway injuries

Anthrax

- Vigorous national effort to develop, manufacture, and stockpile an improved vaccine

Smallpox

- Vigorous national effort to develop, manufacture, and stockpile an improved vaccine
- Major program to develop new antismallpox drugs for therapy and/or prophylaxis

Botulinum Toxins

- Recombinant vaccines, monoclonal antibodies, and antibody fragments

Non-specific Defenses Against Biological Agents

- New specific and broad-spectrum antibacterial and antiviral compounds

Moderate Priority

Nerve Agents

- Intravenous or aerosol delivery of antidotes vs. intramuscular injection
- Development of new, more effective anticonvulsants for autoinjector applications

Cyanide

- Dicobalt ethylene diamine tetraacetic acid, 4-dimethylaminophenol, and various aminophenones
- Antidote stockpiling and distribution system
- Risks and benefits of methemoglobin forming agents, hydroxocobalamin, and stroma free methemoglobin

Phosgene

- N-acetylcysteine and systemic antioxidant effects

Viral Encephalitides

- Antiviral drugs

Viral Hemorrhagic Fevers

- Antiviral drugs

Botulinum Toxins

- Botulinum immune globulin

Low Priority

Brucellosis

- Vaccine

Pneumonic Plague

- Second generation vaccine

Q Fever

- Genes and gene products involved in pathogenesis

Staphylococcal Enterotoxin B (SEB)

- Characterization if mechanism of action
- Active immunization

Ricin

- Antiricin antibodies and formalin treated toxoid immunization

T-2 Mycotoxin

- Screening antivesicant treatments in animal models

9

Prevention, Assessment, and Treatment of Psychological Effects

Incidents of chemical and biological terrorism may involve large numbers of individuals, across all age groups and in both sexes. The survivors of and responders to such incidents will not only suffer physical injury requiring decontamination and medical care but also will undoubtedly undergo extreme psychological trauma. Thus, chemical or biological weapons of mass destruction could produce both acute and chronic psychiatric problems. Unlike storms or floods, chemical disasters occur with little or no warning and are accompanied by continuing fears of ongoing illness and premature death (Bowler et al., 1997) as well as worries about possible genetic or congenital birth defects in subsequent offspring. In the case of terrorism, particularly when the aggressor is unknown, a potentially beneficial expression of anger cannot be directed at the appropriate source, producing a futile sense of helplessness, depression, demoralization, and hopelessness.

LONG-TERM EFFECTS OF TERRORISM (POST TRAUMATIC STRESS DISORDER)

The literature on civilian terrorist attacks reveals a number of reports of very high rates of Post Traumatic Stress Disorder (PTSD) after such attacks. In a study in France, Abenhaim, Dab, and Salmi (1992) followed 254 survivors of terrorist attacks over a period of five years. These authors report that even years after the attacks, the severely injured had a 30.7 percent prevalence of PTSD, and uninjured victims had a 10.5 percent rate. In two other studies of terrorist attacks (Curran et al., 1990; Weisaeth, 1989) PTSD rates higher than 40 percent are reported.

In addition to PTSD, many of the victims of a terrorist attack may suffer the death of family members, close friends, or work colleagues, which can lead to a complicated bereavement with its own elevated risk for depression, self medication, and substance abuse. Many studies indicate that depression is a common co-morbid condition with PTSD. Somatic sequelae to anxiety-related reactions have been reported in most studies of PTSD as well as following the Persian Gulf war. Carmeli, Liberman, and Mevorach (1994) reported that American veterans had a 38 percent prevalence rate of somatic symptoms, and Deahl et al. (1994) report a 50 percent prevalence of some "psychological disturbance suggestive of PTSD" in British soldiers who handled and identified dead bodies of allied and enemy soldiers during the recent Gulf War. These reports suggest that chemical and biological terrorist attacks might cause high rates of PTSD and risks for physical illnesses and suicide, not only among rescue workers but especially among unprepared witnesses to grotesque sights and untrained "good samaritans" voluntarily joining rescue and first aid efforts.

The early identification of persons at risk for long-term psychological effects is complicated by the fact that PTSD symptoms within a few days of a traumatic event have been shown to have low predictive validity by themselves for later psychiatric outcome (Shalev, 1992). Recording of signs and symptoms in the immediate aftermath of the traumatic event should certainly be supplemented by systematic recording of objective and subjective features of the terrorist attack and its aftermath by all who were at the scene. The latter sort of information has often been critical to *post hoc* "prediction" of long-term dysfunction. PTSD is difficult to treat, and even when treated shortly after onset, as was the case with the Japanese sarin victims, 30 percent of the patients required ongoing therapeutic treatment (Nakano, 1995). In addition to the need for rapid identification of those who may require immediate or long-term psychiatric treatment, neuropsychological testing is important to evaluating effects on cognition, memory, and personality as well as any possible organic sequelae from the chemical agents used in terrorist attack.

SHORT-TERM EFFECTS OF TERRORISM (ACUTE NEEDS)

At the acute stage of the aftermath of a biological or chemical terrorist attack, acute autonomic arousal and panic may result in both the victims and the emergency responders (Hazmat teams, police, fire, medical) incapacitating the assistance infrastructure. The severity of these anticipated psychological responses highlights the urgent need for concrete mental health support at times of chemical or biological terrorist attacks.

At the regional and national levels, the American Red Cross Disaster Mental Health Services provides emergency and preventive mental health services to both people affected by the disaster and to Red Cross workers assigned to the disaster relief operation. These services include practical measures like meeting families traveling to the scene, communicating with families not at the scene, offering education about stress and coping, and providing information about local mental health resources. It should be noted in this context that, as victims of a crime, many survivors of a terrorist attack are eligible for compensation and assistance through state victim assistance programs, and, as terrorist attacks are often directed at government buildings, Workman's Compensation. Coordinating access to such sources of financial assistance can be important mental health support. Victims of terrorist attacks are often witnesses in criminal proceedings as well, a role that can change the course of recovery, and may need continuing help in meeting this societal obligation.

The federal government's National Disaster Medical System (NDMS) includes Disaster Medical Assistance Teams with a focus on mental health. Another federal program is the Crisis Counseling Assistance and Training Program (CCP). Funded by the Federal Emergency Management Agency (FEMA) and administered by the Center for Mental Health Services (CMHS) in the Substance Abuse and Mental Health Services Administration, CCP provides supplemental funding to states for short-term crisis counseling services to victims of major disasters. These services are designed to help disaster survivors recognize typical reactions and emotions that occur following a disaster and to regain control over themselves and their environment. Although the focus is on short-term interventions, helping people with normal reactions to abnormal experiences rather than long-term therapy for pathological conditions, the program provides for up to 12 months of services, and local mental health workers and other disaster workers are eligible for training (training is also offered annually to state mental health authorities by FEMA's Emergency Management Institute). Thousands of people were helped by the CCP after the Northridge, California earthquake in 1993, the Oklahoma City bombing in 1995, and the floods in the Dakotas and Minnesota in 1997. CMHS also provides training and field support for a cadre of FEMA employees who provide stress management services to disaster workers.

FIRST RESPONDERS

Another important aspect of terrorism is the emotional and psychological impact on first responders. Prevention methods should be developed to assist first responders in terrorist attacks, whose emotional vulnerability to the traumatic events they face is already recognized in current training programs. Although the U.S. Anti-Terrorism and Effective Death Penalty Act of 1996 recommended a two-day training provided by the DoD Domestic Preparedness Program in 120 targeted urban areas—and, in fact, some 6,500 fire and emergency personnel have been trained within its first year—scant emphasis is paid to the mental health needs of first responders. The current training programs include primarily technical information on the nature and effects of weapons of mass destruction. The focus is on the handling of victims with little attention to the first responders' own mental health and coping needs and strategies. An epidemiologic study conducted by the University of Oklahoma found that 20 percent of the rescue personnel at the Oklahoma City bombing required mental health treatment (Flynn, 1996). This prevalence of mental health problems in first responders demonstrates the need for qualified mental health professionals who can identify and treat vulnerable first responders and so diminish the high rate of mental disorders following terrorist attacks.

Not only is there a need to more effectively identify the mental health needs of first responders, but there is also a need for further research on various treatment methods. For example, Critical Incident Stress Debriefing (CISD), a technique aimed at helping field rescue personnel cope with the stress of extraordinary traumatic events, has gained widespread popularity (Mitchell and Everly, 1996). CISD was originally devised as a relatively rapid technique designed to alleviate stress symptoms and prevent burnout of rescue workers. It involves organized group meetings for all personnel in the rescue unit, with or without symptoms, emphasizes peer support, and is led by a combination of unit members and mental health professionals. CISD in some form has gained wide acceptance among field emergency workers and is increasingly used with hospital-based emergency personnel, military service members, public safety personnel, volunteers, victims, witnesses, and even schoolmates of victims. It can reasonably be expected that many local police, fire, and emergency medical units will be familiar with the CISD process, have access to trained debriefers, and plans for their use. The Metropolitan Medical Strike Teams being organized by the Public Health Service include CISD as part of their standard operating procedures. Objective evidence of CISD effectiveness is, nevertheless, limited and contentious (see for example Raphael et al., 1995; Kenardy et al., 1996; Hamling, 1997), and protocols typically do not allow collecting the type of screening data necessary for estimating psychiatric risk and planning extended services.

NEUROLOGICAL VS. PSYCHOLOGICAL RESPONSES

Although most, if not all, hospitals will have behavioral health staff (psychiatrists, psychologists, psychiatric nurses, counselors, and social workers) present or on call, their experience with PTSD, large-scale disasters, and terrorist acts is likely to be highly variable, and accurate information on chemical or biological agents will be very rare, at least initially. Such information will nevertheless be critical for differentiating those suffering psychological effects from those with neurological damage (Vyner, 1988).

Shapira et al. (1994) suggest an outline of a hospital organization that includes mental health professionals for a chemical warfare attack. Because many of the neurological effects of chemical agents may be confused with the emotional and psychological effects, the authors caution against assuming that the symptoms of chemical trauma victims are psychological in nature and recommend treating victims in a site other than the department of psychiatry.

In nerve gas attacks where the enzyme acetylcholinesterase is inhibited, signs of central and peripheral nervous system poisoning include apathy, mood liability, thought disorders, sleep disorders, and delusions and hallucinations, in addition to psychological stress sequelae. Mental health staff will need to rapidly identify and differentiate the diagnostic characteristics in order to refer and treat these victims as well as rescue workers, who may also suffer from emotional exhaustion and overload.

Important data on this problem of differentiating neurological and psychological effects may emerge from a follow-up study of the 1995 release of sarin in the Tokyo subway (Nakano, 1995). A small number ($N = 34$) of the firefighters who responded are still being followed and are being compared to 36 age- and sex-matched controls recruited from the Tokyo Metropolitan Fire Department. This long-term follow-up study is gathering information on physical and psychiatric sequelae to a nerve gas attack and includes neurological and neuropsychiatric assessments, EKG, peripheral nerve conduction, auditory-evoked potentials, serum cholinesterase levels, and ophthalmic evaluation. Although this follow-up study is of a relatively small number of victims, the information obtained may be useful in planning specific programs for those who may suffer psychological or neurological effects in future terrorist attacks.

TREATMENT METHODS

Because most of the existing studies of PTSD following chemical incidents are predominantly epidemiological in nature, the focus of research has been on sequelae rather than treatment methods and their efficacy.

Treatment is now a very active area of clinical research, with positive treatment outcomes even among torture victims and prospects for improved intervention strategies. However, there is a need to research individual patients' responses to current treatment methods to ensure that favorable outcomes are possible for everyone. Psychological treatment of trauma victims is always complicated by an ongoing need for medical treatment and physical rehabilitation, and in the case of terrorist attacks on civilians, it is further complicated by the fact that attacks frequently occur in everyday settings to which the victims are likely to return. A public approach to treatment should thus include education about the role of reminders and environmental changes that minimize unnecessary and avoidable secondary trauma.

Populations at special risk, such as families with young children, the frail elderly and disabled, may require additional services. Recent research has suggested that the potentially profound trauma reactions in victims should not be treated by publicly venting these fears, as is encouraged in Critical Incident Stress Debriefing (CISD), but instead should receive basic medical and social services, including therapeutic validation of their fears and reassurance that they are reacting normally to an abnormal event (Bisson and Deahl, 1994). That should serve to restore some basic level of daily functioning and to help to restore the needed belief and trust in government institutions.

TRAINING

Current hospital and professional response capabilities should be reviewed for current knowledge about chemical and biological warfare agents so that these personnel will be better prepared to address the emotional sequelae likely to follow an attack with such agents. While there is some knowledge about the psychological effects of terrorism (the Oklahoma City disaster, for example) and of unintentional chemical or biological disasters, little is known about the psychological effects that are specific to chemical or biological terrorism. This lack of knowledge further emphasizes the need for updating protocols and providing additional training of health providers to assure adequate mental health support in existing disaster networks.

Agencies involved in these training effort are: (1) FEMA, providing grants to states, training emergency response, and coordinating delivery of federal counterterrorism; (2) the EPA, whose missions in terrorist attacks are to coordinate personnel and equipment; to respond to hazardous substances, to monitor and to assess the health and environmental impacts; to plan for control, restoration, and disposal of hazardous materials, and to train federal, state, and local response personnel and other responders dealing with hazardous materials emergencies; (3) the Department of Energy and the Nuclear Regulatory Commission, which are charged with providing nuclear and radiological training components to existing emergency response plans; (4) the Department of Health and Human Services, which is the lead federal agency for health and medical services during a presidentially declared disaster; and (5) the Office of Emergency Preparedness, which coordinates the federal health and medical response and recovery activities.

Large cities, such as New York, Chicago, Denver, Boston, and smaller municipalities in Kentucky, Rhode Island, and Massachusetts have developed training programs for their existing emergency response personnel. However, these programs, essential in the immediate remediation of the physical hazards of a terrorist attack, are not currently designed to integrate, plan, provide, or coordinate their efforts with specialty mental health response teams.

The American Psychological Association (APA) has a Disaster Response Network (DRN) of 1,500 psychologists, who have volunteered to provide on-site mental health services to disaster survivors and responders. The DRN services are integrated with the American Red Cross disaster response service and emphasizes brief crisis intervention, primarily for natural disasters. DRN has few psychologists with knowledge of chemical and biological agents that might be used in terrorist attacks. The DRN also does not include education on the neuropsychological impact and need for brief neuropsychological screening of victims of chemical, nerve gas, or biological weapons of mass destruction. DRN does not address either the issue of predictable mass mental health casualties following terrorist attacks or training in the necessary triage and collaboration with the many different state and federal agencies providing services in such an event.

The American Psychiatric Association has a similar but smaller group of volunteers in its Committee on the Psychiatric Dimensions of Disaster. The committee sponsors an introductory course on psychiatric aspects of disasters and a disaster workshop at the Annual Meeting and has developed written and audiovisual materials and distributed them to each of the Association's 77 district branches. In addition to this national level committee, many district branches have disaster committees which respond to local issues and needs. A summary of a 1996 conference on the role of psychiatrists in disasters, jointly sponsored with the American Red Cross and posted on the Internet at www. psych.org, suggested a number of initiatives aimed at educating psychiatrists about the Red Cross Disaster Mental Health Services and fostering a partnership between the two organizations.

In the United States, licensed psychologists, counselors, and psychiatrists are required to complete ongoing professional continuing education. Specific training programs awarding continuing education credits, and possible certification, for mental health crisis intervention after terrorist attacks could be developed and provided by the corresponding major professional organizations.

COMMUNITY EFFECTS

In addition to meeting the psychological needs of individuals, emergency management officials must deal with the reactions of the community as a whole. An important part of any large-scale threat to public health is the psychological effects it engenders in the general public. This will be especially important in the case of chemical or biological terrorism, one goal of which is often to produce fear, panic, demoralization, and loss of confidence in government. Little is known about the fears and feelings engendered by the threat of infection, but considerable research on risk perception and risk communication has been conducted in connection with hazardous waste sites, nuclear power plants, and other real and perceived environmental threats, and general guidelines for government officials have been produced (Hance et al., 1988; National Research Council, 1989; Stern and Fineberg, 1996). Timely provision of accurate information about the nature of the threat and the action being taken to combat it is a central tenet of this advice. Training being provided to major cities through the Army's Domestic Preparedness Program could make that information available if it is implemented as planned (approximately 40 of the 120 cities scheduled had received training by November 1998). The committee has not been able to determine whether these or other cities have prepared information packages of their own for use in informing the press and the public in the event of a terrorism incident, but such preparation will surely be necessary if public officials are to maintain the confidence of a community deluged with information of widely varying accuracy in the news media and, increasingly, on the Internet.

R&D NEEDS

The committee is concerned about several areas of the psychological response to chemical or biological terrorism. Among these areas of concern are the training of mental health professionals, methods for screening victims, and communication to the general public. Therefore, the committee has identified the following research and development needs.

9-1 *Identify resource material on chemical/biological agents, stress reduction after other traumas, and disaster response services, and enlist the help of mental health professional societies in developing a training program for mental health professionals. The key to success in this attempt will be offering continuing education credits and certification for mental health providers trained in chem/ bio attack response.*

9-2 *Identify suitable psychological screening methods for use by mental health providers and possibly first responders, differentiating adjustment reactions after chem/bio attacks from more serious psychological illness (e.g., panic disorder, PTSD, psychosis, depression), and organic brain impairment from chemical or biological agents. Research to identify trauma characteristics and behavior patterns that predict long-term disability may be necessary.*

9-3 *Develop health education and crisis response materials for the general public, including specific communication on chemical or biological agents. Additional information is needed on risk assessment/threat perception by individuals and groups and on risk communication by public officials, especially the roles of both the mass media and the Internet in the transmission of anxiety (or confidence). Some information is available in EPA studies of pollutants and toxic waste, but there is little or no systematically collected data on fears and anxieties related to the possibility of purposefully introduced disease.*

9-4 *Evaluative research is needed on interventions for preventing or ameliorating adverse psychological effects in emergency workers, victims, and near-victims. Specific crisis intervention methods may be necessary for chemical or biological terrorist incidents, but in the absence of such incidents researchers might draw on studies of chemical spills, epidemics of infectious disease, and more conventional terrorist incidents.*

10

Computer-Related Tools for Training and Operations

Those responsible for providing the medical response to a chemical or biological terrorist attack on a civilian population will face extraordinary crisis control and consequence management problems. Depending on the specific nature of the event (e.g., threatened or actual, release of chemical or biological substances), these first responders may have to (1) immediately provide and coordinate adequate first-aid and critical emergency medical assistance; (2) identify prospectively or retrospectively the location, type (chemical and/or biological), and mechanism of release, such as a stationary or mobile spray, an explosive device, etc. (the "source-term" in computer models), and construct a reasonable footprint for exposure (e.g., atmospheric dispersion over space and time) and potential doses; (3) conduct a hazard assessment and recommend practical intervention procedures (e.g., isolation, shielding, distribution of pharmaceuticals) to limit exposures and further amplification of adverse health effects; and (4) determine the extent of physical contamination and then isolate and decontaminate property to restore and salvage landscapes, buildings, and transportation for rapid reutilization. In parallel with these tasks, and perhaps even competing and conflicting with them, will be separate efforts devoted to collecting and preserving evidence in order to apprehend and prosecute the perpetrators. Therefore, the most effective action will depend on all first responders communicating and coordinating their actions, as well as working closely with federal, state, and local authorities, healthcare institutions, and even news services. Clearly, for civilian medical and law-enforcement first responders to address these acts of terrorism optimally and rapidly, their collective efforts will need some choreographing, and they will have to react instinctively and collaboratively as they do in other emergency situations for which they have been adequately trained.

Fortunately, medical and other first responders can acquire these essential instinctive and collaborative reactions for responding to an actual or threatened chemical or biological terrorist act by enhancing their existing skills, knowledge, and abilities for dealing with more conventional disasters. However, as unlikely a chemical or biological terrorist act is in any given locale, its potential impact makes it vital to the first-responder community, especially the principal decisionmakers, that such enhancement of existing capabilities also be sustained. Accordingly, this section of the report identifies relevant computer-related tools and pertinent health-effects information that could be used by medical and other first responders to train regularly or even use operationally. These tools will also decrease the need for frequent participation in large exercises that can be disruptive, logistically complicated, expensive, and unproductive.

MEDICAL VIGILANCE AND DOSE RECONSTRUCTION

Extremely rare infections, chemical exposures, or alternatively, temporally or geographically unusual or uncommonly frequent adverse-health effects could serve as an early warning that there has been a covert release of a chemical or biological substance into a civilian population. Emergency care facilities are likely to be the sentinels for observing such effects in a population. Consequently, the medical community can actively contribute to the rapid identification of a chemical or biological release if they have at their disposal communication systems by which they easily can report confirmed or suspected, rare diagnoses to public health officials.

As mentioned in Chapter 5, *Recognizing Covert Exposure in a Population,* although epidemiological surveillance systems exist and public health authorities do compile some health-effects information (e.g., morbidity/mortality reports), the process is slow, somewhat isolated, and should be better networked so that data streams documenting rare events can be received, assimilated, and analyzed for trends far more rapidly. In fact, a computer network, combined with easily understood software, perhaps involving the Internet, and an approach that is similar to or is connected with the Program for Monitoring Emerging Diseases (ProMED) or the Global Infectious Disease and Epidemiology Network (GIDEON) could be designed for rapidly collecting diagnostic data from the medical community electronically, particularly from sentinel locations, such as emergency departments. These data could then be sent to a secure, centralized electronic data collection point for compilation, prompt assessment, and distribution of results, along with the raw data, to local, state, and national levels for further analyses.

Although the centralized system and its analytical tools are not currently available, the benefits of developing rapid assessment procedures for addressing the accumulating data would be extremely valuable. Not only would it contribute to forensic epidemiology related especially to covert acts of biological and perhaps chemical terrorism, but it would also help to recognize instances of emerging disease or infection. For example, a computerized analysis could be designed to promptly detect in the data any unusual disease or chemical toxicity event(s), as well as those with particular characteristics related to specific chemicals or microorganisms, that might otherwise be ignored or uncorrelated because of infrequency or geospatial and/or temporal dispersion, and then alert public health authorities to this finding.

Should the reported symptomatology for certain individuals signal a possible covert release of chemical or biological substances in a civilian population, the computer system could also aid in determining the environmental media of exposure (e.g., air, water, or food release) and assist public health, law enforcement, and hazardous-material authorities in reconstructing the source, footprint of the exposures, and spectrum of doses. It could do this by modeling the applicable vectors of dispersion (e.g., wind speed and direction, water flows, transportation systems, or even the distribution of contaminated goods or services). Swift retrospective analyses of these data would enable public health authorities to more quickly deduce the source, isolate areas of exposure, locate contaminated property, and distribute available vaccines and antidotes and beneficial information. Also, law enforcement personnel could act to acquire evidence to catch the perpetrator, evidence that might otherwise become undetectable over time. The benefits of having such a system are clearly demonstrated by Meselson et al. (1994). In this case, the investigators effectively combined medical, biological, meteorological, and demographic data to demonstrate retrospectively and convincingly that atmospherically released anthrax from a Russian production facility was responsible for incidents of infection in the downwind community of Sverdlovsk in 1979.

MODELS FACILITATING ASSESSMENT AND PLANNING

Even if individual first responders can be adequately equipped and prepared to safely handle the physical and emotional hazards of attending to victims and their families following an act of chemical or biological terrorism, it is crucial that the efforts of these individuals be administered and coordinated systematically and objectively. Such direction is necessary for minimizing or eliminating additional exposures, averting needless pain and suffering, and preventing any amplification of serious adverse health effects in the population or among first responders.

An important part of the task of providing the guidance needed by the first-responder community to react rapidly, competently, collaboratively, and instinctively during and after a chemical or biological terrorist event is to provide a mechanism for establishing a clear understanding of how to quickly adjust to different environmental circumstances (e.g., meteorological conditions, hydrological events, and geophysical formations), human-behavior (e.g., traffic and mass hysteria), infrastructure limitations (e.g., availability of hospitals, pharmaceuticals, and services) and communication interruptions (e.g., network breakdowns). Field exercises are one such mechanism, but these seldom can address more than one issue at a time, may be difficult to conduct frequently due to scheduling constraints by participants, and may be expensive due to the required levels of personnel and equipment involved. An alternative is offered by advances in computers and software, which make it possible to address the essential training and operational requirements more conveniently and cost-effectively. In fact, such computer-related tools could also help formulate responses to unintended releases of conventional chemicals and hazardous materials. Thus, if developed, such computer-related tools could permit first-responders to enhance and sustain their ability to assess and plan for a variety of different situations. The computer models that could be used now and in the future are discussed next, along with the importance of an improved understanding of the toxicological properties of the chemical and biological substances that might be used for acts of terrorism.

During any threatened or actual act of chemical or biological terrorism, the immediate reaction of the first-responder community will be to identify the specific agent, determine the best methods for reaching and treating any exposed individuals, decide whether to evacuate any critically ill or other potentially susceptible members of the population (e.g., children, seniors, etc.), and consider the most appropriate ways to avert further exposures and casualties. These efforts will require an understanding of chemical and physical properties of the agent, its likely mechanism and location of release, its environmental transport and fate, and the acute and chronic health effects resulting from both low and high dose levels. This can involve individual or multiple exposure pathways (e.g., across skin, lungs, or gastrointestinal tract) and translate into a variety of symptoms, some that may not even require medical intervention.

As described in Chapters 4 and 6, which address detection and measurement of chemical and biological agents, research is under way in analytical chemistry and genomics to provide advanced techniques and miniature devices that rapidly and accurately detect and recognize small concentrations of chemicals and microorganisms present in environmental and biological samples. First responders would then have at their disposal analytical devices for more rapidly determining the presence of a chemical or biological substance in a sampled environmental media or biological fluid.

Once the agent is known, its toxicological and chemical properties will be of interest to those responding to the incident. Currently, this information can be obtained by verbal communication with local poison control centers and/or experts in the federal government, but eventually the information and experts might be even more promptly available by computer network (see Chapter 5). Much of the currently available toxicological information is not comprehensive for the chemical and biological substances considered in this report, and in most cases documents only lethal dosage or acute effects by a specific exposure pathway. It is nevertheless reasonable to expect even such limited information will be used and extrapolated, if necessary, pending the development of more precise and relevant data. In fact, such a conclusion about the quality of the toxicological information available is consistent with results contained in a review of the acute-human toxicity estimates for selected chemical-warfare agents performed by the Committee on Toxicology of the National Research Council (COT/NRC, 1997). Developing more information to address the toxicological behavior of such substances, including physiologically based pharmacokinetic (PBPK) models for estimating biochemical metabolism, is necessary for understanding the full range of health effects likely to be seen during and after a release, especially those likely to occur from low-dose exposures and that might not require extensive medical intervention. Furthermore, such information is valuable for providing realistic instructions for using protective equipment and for authorizing reentry into contaminated areas and the decontamination of property.

Where public event planning requires that consideration be given to preparing for chemical or biological terrorism, or there is advance knowledge of the likely location, timing, and type of such a terrorist act, then transport and fate modeling can be performed to determine the extent of the release and to identify the population likely to be exposed. The models that would be used for this purpose are those that assess the movement and dissipation of the agent and identify potential locations of serious exposures as well as surfaces of contamination. Combined with dose-response algorithms and demographic data, these models also can be used to translate concentrations in environmental media (e.g., air, water, and soil) into casualty maps.

Scenarios for terrorist acts that involve the release of chemical or biological substances into ambient air are considered among the most likely, as this is an easy way for a terrorist to achieve dispersion, affect a large population, and gain attention. Therefore, models currently available or that are undergoing development primarily focus on identifying the consequences of a release into air.

For addressing release directly into the atmosphere the available models range from the simple (Gaussian puff simulating advection and dispersion) that can be operated on a desktop personal computer to the complex (three-dimensional, particle-tracking models that use real-time acquisition of local meteorological data and account for terrain). The latter models are computationally intensive and require larger computers and specialists for their operation. However, to be applied correctly, even desk-top computer programs at this time are technologies that require the user possess a good degree of familiarity with the software and its operation and a reasonable knowledge of the model attributes and limitations. Mazzola et al. (1995) describe many of the different atmospheric dispersion models currently available and indicate whether they are governmental or commercial.

There also are other more specialized models for describing the behavior of materials released into the airflow near buildings and into particular structures (e.g., buildings and subways). A description of the purpose of these models appears in a more recent U.S. Department of Energy (DoE) publication to which several National Laboratory research groups contributed (ANL, LBNL, LLNL, and LANL, 1997). That document focuses on explaining recent developments and plans by DoE researchers to direct their atmospheric-science and computer-simulation expertise toward improving transport and fate modeling for application to urban environments, building interiors, and subway systems. This becomes necessary because the accuracy in predicting the transport and fate of material released above urban terrain, inside buildings, and within subway systems of metropolitan areas requires more scaling (e.g., finer grid resolution) and physical considerations (e.g., more complex fluid dynamics) than do currently available regional-scale diagnostic models.

In applying any atmospheric-dispersion model, it is important that the source-term properties be reasonably well defined. Some atmospheric-dispersion models now available, for example, VLSTRACK (Bauer and Gibbs, 1996), have attempted to address specifically the likely methods used to release a chemical or biological substance and then attempt to describe adequately the resulting dispersive and advective nature of the release. Models of this type have in common that they can predict concentrations in air, and to some degree, the footprint of deposition during and after release. However, the current versions of such models address specific physical processes, landscapes, and even sources of release, and one model may provide better results than another, depending on the situation being considered. Furthermore, requirements for model selection, including sophistication, accuracy, and computer power, may very well depend not only on how the model addresses the type and nature of the release, but also the degree to which the model can approximate the terrain over or through which the released material will disperse (e.g., simple or complex, rural or urban), and the quantity of the meteorological information that is needed and can be made available or approximated.

Along with VLSTRACK there are other atmospheric-dispersion modeling programs. Some are designed to combine atmospheric-dispersion modeling with effects analyses to help emergency responders and decisionmakers address chemical releases. Among these models are CAMEO (computer-aided management of emergency operations), which uses ALOHA (areal locations of hazardous atmospheres) as its atmospheric-dispersion model. This model was developed by the U.S. Environmental Protection Agency and the National Oceanic and Atmospheric Administration (1996) to assess unintentional chemical releases. The program operates on a personal computer, but would require some adaptation (e.g., specific information about the type of material released and the physical characteristics of the source-term) to address the specific release of a chemical or biological agent into an urban environment. Another modeling system for assessing potential hazards specifically related to the release of particular chemical, biological, and nuclear materials is HPAC (hazard prediction and assessment capability), which includes the Second-order Closure Integrated Puff (SciPUFF) model for assessment of atmospheric transport. This software also operates on a personal computer and is distributed by the Defense Special Weapons Agency (1997) to aid in hazard assessment relating to the atmospheric transport of a chemical, biological, or nuclear source-term that can be estimated. The adequacy of any one or all of these models for application by first-responders requires further examination, including evaluation of their applicability to different situations and requirements for operational expertise.

Finally, one of the most sophisticated computer simulation programs for hazard assessment is the property of the DoE and is located in California at the Lawrence Livermore National Laboratory's National Atmospheric Release and Advisory Center (NARAC). The Center's primary responsibility involves predicting the dispersion of accidentally released radioactive materials, but the system can address a variety of other substances as well. Because this system uses real-time meteorological information, particle tracking, and accounts for complexity of terrain, it is a numerically complex tool that does not operate on a desktop personal computer and requires trained personnel for operation and interpretation. The more information that can be provided about the source-term and meteorology, the more accurate the predictions of dispersion and effect will be and the faster such information can be obtained and fed back to the requester.

If any of the models just discussed are to be employed during a real emergency, it must be emphasized that their operation and limitations must be familiar to the users. In fact, NARAC represents a valuable tool to the DoE precisely because it is a dedicated operation and its operators practice regularly and can be called upon any time of day. However, there is a cost associated with its operation, because of its centralization, dedicated personnel, and extensive data requirements. Nevertheless, the system employed by NARAC personnel, with recent enhancements specifically addressing release of a chemical or biological agent, has been successfully employed to plan protection of the public at events of special significance (Ermak, 1998). Similarly, recent modifications to subway ventilation models (Policastro, 1998) and to indoor air models (Sextro, 1998) have made it possible to apply these models to release of chemical or biological agents, although at this time only under certain conditions and primarily by the model developers. It is conceivable that emergency response organizations might solicit the use of these models to plan appropriate responses to releases of chemical or biological substances during special events. Unfortunately, such models may be too difficult or too costly for many communities to take advantage of. In such cases, it may be more appropriate that they independently develop their own expertise in using less sophisticated desktop computer models. With such expertise, it is conceivable these models could be used to produce contingency plans based on conservative parameter estimates, as well as provide conservative estimates of concentration levels over a landscape in the event of real emergencies.

Training to respond to a chemical or biological terrorist act can involve using any currently available atmospheric-dispersion modeling system and would help prepare the first responder community intellectually to deal with unfamiliar situations. However, current computer models for addressing the transport and fate of substances are not interactive and do not reflect to any significant degree the movements of people temporally or spatially during and after the event. To address this situation there is currently an effort under way at the Lawrence Livermore National Laboratory to introduce into a conflict-simulation software system realistic scenarios involving releases of chemical and biological substances. Specifically, the conflict-simulation software has been used successfully as a training tool for military actions on a battlefield (Sackett, 1996), and is undergoing proof-of-principle modification to include results from an atmospheric-dispersion program and relevant toxicological data, so it can also address health consequences of chemical and biological releases. In the conflict simulation, multiple operators are allowed to role play interactively and move their personnel and equipment during the simulated incident. For example, the simulation software allows individuals and groups to be moved on a variety of urban landscapes and in the presence of different meteorological conditions (e.g., rain, wind, sunshine) and human activities (e.g., traffic patterns, mass exodus, and the actions of perpetrators and responders), all during a specifically designed scenario involving passage of an atmospherically dispersed substance. The dose-response algorithms that are introduced permit the model to provide information about declining performance or death likely to be observed in exposed individuals, both stationary and moving in and out of the cloud. This conflict-simulation model allows communication failures, changing weather conditions, and human-behavior patterns to be introduced into the scenario, giving users the opportunity to respond to situations changing dynamically and to immediately visualize the results of their acts of commission and omission.

SUPPORT FOR DECONTAMINATION AND REOCCUPATION STRATEGIES

In many cases unexposed and decontaminated populations are going to need access to safe zones and routes of evacuation. Additionally, contaminated people, structures, and landscapes are going to require cleanup. Certifying exposed facilities, structures, and vehicles as suitable for reuse and individuals as being adequately decontaminated requires defining safe levels for released substances and their degradation products. Performing quantitative health-risk assessments for such substances would generate such information. Such assessments would benefit greatly from, and represent another reason for obtaining, a better understanding of the toxicological and chemical behavior of the substances that could be released in an act of terrorism.

R&D NEEDS

The Committee advocates the following research and development efforts be undertaken to enhance and sustain the capabilities of the medical community to deal with chemical and biological terrorism. Such events, serious as they are, have a low probability of occurrence, but the products of these R&D efforts will also help to identify emerging infections and diseases and to respond to events involving hazardous substances released unintentionally in industrial settings.

10-1 *Develop rapid assessment procedures that include facilitating the reporting, collection, evaluation and distribution to public health authorities of unusual medical symptomatology and their origins. Such a system should be designed to link medical vigilance in civilian populations to computer networks that can capture and evaluate the data quickly, using appropriate models for vectors of dispersion (e.g., meteorological data addressing likely wind speed and direction at suspected time of incident, or parameters related to transportation systems) and available toxicological information for suspected substances. This system would help to reveal likely sources of covert chemical or biological terrorism, and to recognize origins of emerging disease or infection in a timely manner.*

10-2 *Examine current atmospheric-dispersion models and those under development to determine which would be most suitable for the emergency management community for understanding the consequences of a release in air of a chemical or biological substance in an act of terrorism. This evaluation would also help to determine whether it is more appropriate and cost-effective to support large, complex, centralized modeling systems with dedicated operators or to encourage individual communities to recruit and train individuals to run distributed, desktop software models that could both support planning and be used during an actual event. Additionally, research should be conducted to produce computer-related methods for prompt modeling of the other possible vectors of dispersion (e.g., water, food, and transportation) in which chemical and biological substances can be released and transported.*

10-3 *Generate and support interactive simulation software at the national level in order to prepare first-responders and other emergency management personnel for acts of chemical or biological terrorism. Such systems represent a training tool that can be designed to be user friendly, easy to learn, run on networks that can be accessed at multiple locations, and used frequently by all levels of the first-responder community. Additionally, these models can be customized to meet the needs of individual communities, and will reduce the costs and inconveniences associated with staging frequent exercises while permitting the emergency management community to enhance and sustain capabilities to realistically plan for and adjust to unanticipated environmental changes, communication failures, and human behavior.*

10-4 *Conduct research to better understand the chemical, physical, and toxicological properties of the chemical and biological substances that could be employed in acts of terrorism. Such information would improve modeling of their environmental transport and fate as well as their mechanism for producing acute and chronic health effects from both low and high dose levels. This information will also support the health-risk assessments that are needed to make recommendations for performing decontamination and allowing reoccupation of buildings, vehicles, and landscapes following an event.*

11

Conclusions and Recommendations

The preceding pages have addressed a wide range of issues related to effective medical response to acts of chemical or biological terrorism. Each chapter draws some conclusions about a single aspect of that response and makes some recommendations for desirable research and development. There are, nevertheless, some general conclusions, some stated, some implicit, which pervade the report as a whole. The most basic of these is that terrorist incidents involving biological agents, especially infectious agents, are likely to be very different from those involving chemical agents, and thus demand very different preparation and response. Figures 11-1 and 11-2 illustrate these differences in flow diagrams of actions involved in coping with what the committee views as the most likely chemical (Figure 11-1) and biological (Figure 11-2) terrorism scenarios. The diagrams are descriptive, not prescriptive, and certainly do not represent the only possible sequences of action. We believe they are representative, however, and illustrate the contrast between the relatively linear sequence of actions in the chemical event and the more diffuse, parallel, and recursive activities in the biological event. The myriad of "chemical/biological" response teams being developed at federal, state, and local levels are, despite their names, almost entirely focused on detection, decontamination, and expedient treatment of chemical casualties. For both types of incidents, however, there is an existing response framework within which modifications and enhancements can be incorporated. An attack with chemical agents is similar to the hazardous materials incidents that metropolitan public safety personnel contend with regularly. A

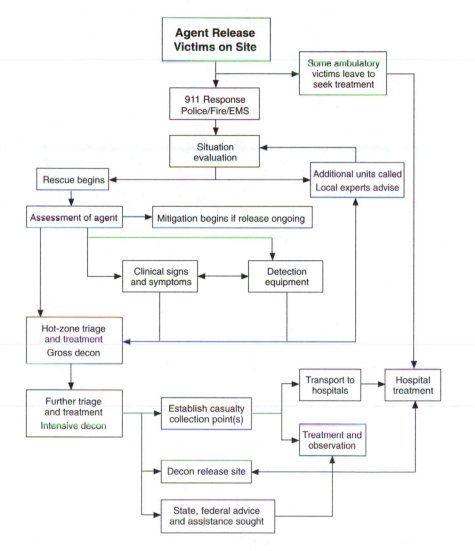

FIGURE 11-1 Flow chart of probable actions in a chemical agent incident.

major mission of public health departments is prompt identification and suppression of infectious disease outbreaks, and poison control centers deal with poisonings from both chemical and biological sources on a daily basis. It would be a serious tactical and strategic mistake to ignore (and possibly undermine) these mechanisms in efforts to improve the response of the medical community to additional, albeit very dangerous, toxic

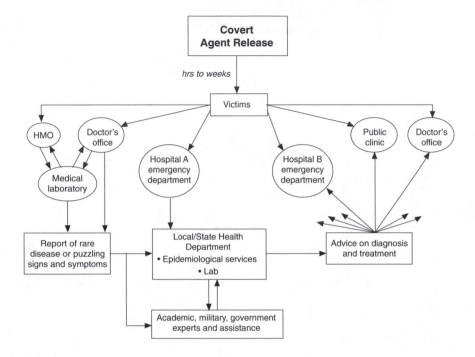

FIGURE 11-2 Flow chart of probable actions in a biological agent incident.

materials. Strengthening existing mechanisms for dealing with unintentional releases of hazardous chemicals, for monitoring food safety, and for detecting and responding to infectious disease outbreaks, is preferable to building a new system focused solely on potentially devastating but low-probability terrorist events. Indeed, a major reason for the committee's decision to focus the report on response to aerosol attacks with the short list of agents thought to be a threat by U.S. military forces was that these agents are unfamiliar to the U.S. civilian medical system. Regardless of relative probability of use or relative lethality, there are mechanisms in place for dealing with a wide variety of other agents and routes. Our concern was not to foster construction of yet another mechanism, but to encourage the incorporation of these unfamiliar agents and routes into existing mechanisms.

A second general conclusion relates to a question which underlays the whole study: whether military approaches to chemical and biological defense are applicable to domestic civilian situations involving these agents. The report points out several aspects of military standard operating procedure that, as the sponsors feared, will be difficult or impossible to implement in a very heterogeneous and independent civilian population. More importantly, the committee was impressed with the extent to which differences in prior knowledge about the identity of the enemy and the time and place of attack lead to important differences in the needs of military and civilian medical communities. Vaccination, for example, is an obvious preventive measure for a military force poised for combat against an enemy known or suspected to have a stockpile of certain biological weapons. The same holds true for deployment of chemical or biological detection systems, the use of highly specific antidotes and therapeutic and pretreatment drugs: with reasonable intelligence about the enemy's capabilities and proclivities, these tools can be put into action rapidly and confidently. The value of all of these actions diminishes considerably in the most probable civilian terrorism situations, in which the enemy, the agent, the time, and the place of attack are unknown. This difference, even more than differences in the physiology of civilian and military targets, influenced the committee to emphasize treatment over prevention, broad-spectrum drugs, detection with familiar or multiagent equipment, clinical diagnosis based on commercial technology, decontamination without agent-specific equipment or solutions, modification of familiar or multipurpose protective clothing and equipment, and even the advisability of prehospital treatment. Chapter 3 argues for including the medical community in the distribution of pre-incident intelligence to maximize medical response in dealing with chemical or biological incidents, but, important as that is, the time scale envisioned in those arguments is much too short for truly preventive measures like vaccination or the introduction of unfamiliar specialized equipment.

A third conclusion which shaped the committee's recommendations concerned problems of scale. In many of the areas surveyed in the previous section, we noted that some capability, often quite good capability, existed for incidents involving a small number of victims. Regardless of preparation, there will be some unpreventable casualties in all but the most incompetent attacks, but without planning, education, supplies, equipment, and training, the casualty count will mount rapidly when the number of persons exposed escalates, particularly as the event is likely to be unprecedented in a community. Local governments and hospitals are reluctant to spend large amounts of money and time preparing for what they judge as low-probability events. Therefore, although the need for integrated planning cannot be overstated, federal organizations can be very important. Because of the rapidity with which chemical agents act, federal help may actually be of less use in a chemical attack, for which they are much better prepared, than in a biological attack, where onset of signs or symptoms is delayed, variable, and potentially continuing, and victims are widely dispersed. The National Disaster Medical System (NDMS), for example, would be a critical component of response to any large-scale biological attack. The NDMS might also serve a useful role in a large-scale chemical attack, though the rapid onset of effects from these agents puts a premium on actions within the first few hours following exposure. For that reason, the Metropolitan Medical Strike Teams being organized and equipped by the Public Health Service may be the most useful federal help in managing the medical consequences of a chemical attack. Similar help from deployable military teams will be optimal only if intelligence allows for predeployment or the attack occurs near the team's home base.

Detection and identification of agents, either in the environment or in victims' bodies, is currently a piecemeal operation that, in the absence of other information, is as much art as science. In both chemical and biological agent incidents, initial treatment of victims is likely to remain symptom-based for some time. In part this is due to diagnosis problems (knowing what detector to deploy in the environment or what medical test to request), limited detection capability at low but potentially harmful concentrations, and lack of specific treatments for some agents. These difficulties are clearly amenable to technological solutions, and the committee is optimistic about the prospects for faster, easier, more specific patient diagnostics. The committee's recommendations on detection and identification of agents in the environment, however, were shaped very strongly by assumptions about terrorism scenarios: that vapor or aerosol delivery will mean that agents may be difficult to locate 10, 20, or 30 minutes after a chemical agent release, when the first detectors arrive at the scene, and that the release site and time of a biological attack will not be known for days or weeks after the release, if at all.

Finally, it is apparent that requirements of federal regulatory agencies (OSHA, FDA) not primarily concerned with emergency response to low-frequency events like chemical or biological terrorism nevertheless have a substantial influence on response capabilities. The characteristics and rules for use of personal protective equipment, for example, fall under the jurisdiction of the Occupational Safety and Health Administration. The investigational (IND) status of some very specific treatments, present and future, will hamper their use in mass-casualty situations. Furthermore, in the case of many treatments, collection of the data on efficacy necessary for full FDA approval will not be possible for ethical reasons or economically attractive to a potential manufacturer because of limited market potential.

RECOMMENDATIONS FOR RESEARCH AND DEVELOPMENT

As expected, the committee's review of current capabilities pointed to a number of areas in which innovative R&D is clearly needed. Detailed, specific lists of R&D needs are offered at the end of each chapter (61 in all), and they are summarized below in the form of 8 overarching recommendations. As the text and the inventory in Appendix B reflect, there is a great deal of relevant R&D under way in both the public and private sectors that may meet some of the needs we point out, and the following list of recommendations should not be construed as commentary on the quality of that research or the utility of its intended products for military applications. The order within the list is not by priority, but follows the roughly chronological order of the chapters of this report.

Recommendation 1. There needs to be a system in every state and major metropolitan area to ensure that medical facilities, including the state epidemiology office, receive information on actual, suspected, and potential terrorist activity.

Specific R&D Needs:

- A formal communication network between the intelligence community and the medical community.
- A national mechanism for the distribution of clinical data to the intelligence and medical communities after an actual event or exercise.

Recommendation 2. The committee endorses continued testing of civilian commercial products for suitability in incidents involving chemical warfare agents, but research is still needed addressing the bulk, weight, and heat stress imposed by current protective suits, developing a powered air respirator with greatly increased protection, and providing detailed guidance for hospitals on dermal and respiratory protection.

Specific R&D Needs:

- Increased protection factors for respirators.
- Protective suits with less bulk, less weight, and less heat stress.
- Evaluation of the impact of occupational regulations governing use of personal protective equipment.
- Uniform testing standards for protective suits for use in chemical agent incidents.
- Guidelines for the selection and use of personal protective equipment in hospitals.
- Alternatives to respirators for use by the general public.

Recommendation 3. The civilian medical community must find ways to adapt the many new and emerging detection technologies to the spectrum of chemical and biological warfare agents. Public safety and rescue personnel, emergency medical personnel, and medical laboratories all need faster, simpler, cheaper, more accurate instrumentation for detecting and identifying a wide spectrum of toxic substances, including, but not limited to military agents, in both the environment and in clinical samples from patients. The committee therefore recommends adopting military products in the short run and supporting basic research necessary to adapt civilian commercial products wherever possible in the long run.

Specific R&D Needs:

* Evaluation of current Hazmat and EMS chemical detection equipment for ability to detect chemical warfare agents.
* Miniaturized and less expensive gas chromatography/mass spectrometry technology for monitoring the environment within fixed medical facilities and patient transport vehicles.
* Standard Operating Procedures for communicating chemical detection information from first responders to Hazmat teams, EMS teams, and hospitals.
* Simple, rapid, and inexpensive methods of determining exposure to chemical agents from clinical samples.
* Faster, cheaper, easier patient diagnostics that include rare potential bioterrorism agents.
* Inexpensive or multipurpose biodetectors for environmental testing and monitoring.
* Basic research on pathogenesis and microbial metabolism.
* Scenario-specific testing of assay and detector performance.

Recommendation 4. Improvements in CDC, state, and local surveillance and epidemiology infrastructure must be undertaken immediately and supported on a long-term basis.

Specific R&D Needs:

* Improvements in CDC, state, and local epidemiology and laboratory capability.
* Educational/training needs of state and local health departments regarding all aspects of a biological or chemical terrorist incident.
* Faster and more complete methods to facilitate access to experts

and electronic disease reporting, from the health care provider level to global surveillance.

• Expanded pathogen "fingerprinting" of microbes likely to be used by terrorists and dissemination of the resulting library to cooperating regional laboratories.

• Symptom-based, automated decision aids that would assist clinicians in the early identification of unusual diseases related to biological and chemical terrorism.

Recommendation 5. R&D in decontamination and triage should concentrate on operations research to identify methods and procedures for triage and rapid, effective, and inexpensive decontamination of large groups of people, equipment, and environments.

Specific R&D Needs:

• The physical layout, equipment, and supply requirements for performing mass decon for ambulatory and nonambulatory patients of all ages and health in the field and in the hospital;

• A standardized patient assessment and triage process for evaluating contaminated patients of all ages;

• Optimal solution(s) for performing patient decon, including decon of mucous membranes and open wounds;

• The benefit vs. the risk of removing patient clothing;

• Effectiveness of removing agent from clothing by showering;

• Showering time necessary to remove chemical agents;

• Whether high-pressure/low-volume or low-pressure/high-volume spray is more effective for patient decontamination;

• The best methodology to employ in determining if a patient is "clean"; and

• The psychological impact of undergoing decontamination on all age groups.

Recommendation 6. Optimize the utilization of currently available antidotes for nerve agents and cyanide though operations research on stockpiling and distribution, and give high priority to research on an effective treatment for vesicant injuries, investigation of new anticonvulsants, and antibody therapy for nerve agents, development of improved vaccines against both anthrax and smallpox, development of a new antismallpox drug, and research on broad spectrum antiviral and novel antibacterial drugs.

BOX 11-1
R&D Needs in Availability, Safety, and Efficacy of Drugs and Other Therapies

HIGH PRIORITY

Nerve Agent
- Antidote stockpiling and distribution system
- Scavenger molecules for pretreatments and immediate post-exposure therapies

Vesicants
- An aggressive screening program focused on repairing or limiting injuries, especially airway injuries

Anthrax
- Vigorous national effort to develop, manufacture, and stockpile an improved vaccine

Smallpox
- Vigorous national effort to develop, manufacture, and stockpile an improved vaccine
- Major program to develop new antismallpox drugs for therapy and/or prophylaxis

Botulinum Toxins
- Recombinant vaccines, monoclonal antibodies, and antibody fragments

Non-specific Defenses Against Biological Agents
- New specific and broad-spectrum anti-bacterial and anti-viral compounds

MODERATE PRIORITY

Nerve Agents
- Intravenous or aerosol delivery of antidotes vs intramuscular injection
- Development of new, more effective anticonvulsants for autoinjector applications

Cyanide
- Dicobalt ethylene diamine tetraacetic acid, 4-dimethylaminophenol, and various aminophenones
- Antidote stockpiling and distribution system
- Risks and benefits of methemoglobin forming agents, hydroxocobalamin, and stroma-free methemoglobin

Phosgene
- N-acetylcysteine and systemic antioxidant effects

Viral Encephalitides and Viral Hemorrhagic Fevers
- Antiviral drugs

Botulinum Toxins
- Botulinum immune globulin

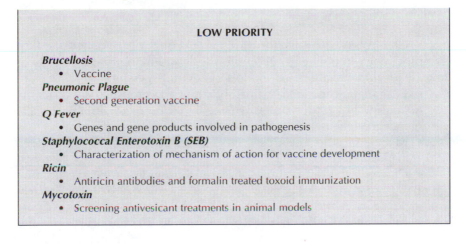

LOW PRIORITY

Brucellosis
- Vaccine

Pneumonic Plague
- Second generation vaccine

Q Fever
- Genes and gene products involved in pathogenesis

Staphylococcal Enterotoxin B (SEB)
- Characterization of mechanism of action for vaccine development

Ricin
- Antiricin antibodies and formalin treated toxoid immunization

Mycotoxin
- Screening antivesicant treatments in animal models

Specific R&D Needs:

- See Box 11-1 for a complete listing by agent and priority.

Recommendation 7. Educational materials on chemical and biological agents are badly needed by both the general public and mental health professionals.

Specific R&D Needs:

- Identify resource material on chemical/biological agents and enlist the help of mental health professional societies in developing a training program for mental health professionals
- Psychological screening methods for differentiating adjustment reactions after chemical/biological attacks from more serious psychological illness.
- Evaluation of techniques for preventing or ameliorating adverse psychological effects in emergency workers, victims, and near-victims.
- Agent-specific information on risk assessment/threat perception by individuals and groups and on risk communication by public officials.

Recommendation 8. The committee recommends support for computer software R&D in three areas: event reconstruction from medical data, dispersion prediction and hazard assessment, and decontamination and reoccupation decisions.

Specific R&D Needs:

• Computer software for rapid reporting of unusual medical symptomology to public-health authorities and linking that data to both toxicological information and models of agent dispersion.

• Examination and field-testing of current and proposed atmospheric-dispersion models to determine which would be most suitable for the emergency management community.

• Models of other possible vectors of dispersion (e.g., water, food, and transportation).

• Customizable simulation software to provide interactive training for all personnel involved in management of chemical or biological terrorism incidents.

• Information on the chemical, physical, and toxicological properties of the chemical and biological agents, in order to improve modeling of their environmental transport and fate and to better support recommendations on decontamination and reoccupation of affected property.

References

Abenhaim L, Dab W, Salmi LR. 1992. Study of civilian victims of terrorist attacks (France, 1982–1987). *Journal of Clinical Epidemiology* 45(2):103–109.

Andersen AH. 1946. Experimental studies on the pharmacology of activated charcoal: I Adsorption power of charcoal in aqueous solutions. *Acta Pharmacologica* 2:69–78.

Andersen GL, Simchok JM, Wilson KH. 1996. Identification of a region of genetic variability among *Bacillus anthracis* strains and related species. *Journal of Bacteriology* 178:377–384.

Anderson GP, Breslin KA, Ligler FS. 1996. Assay development for a portable fiberoptic biosensor. *ASAIO Journal* 42:942–946.

Arad M, Epstein Y, Roybert M, Berkenstadt H, Alpert G, Shemr J. 1994. Physiological assessment of the passive children's hood. In: Danon YL and Shemer J (Eds). *Chemical Warfare Medicine*. New York: Gefen. Pp. 75–80.

Arca VJ, Ramos GA, Reeves DW, Blewett WK, Fatkin DP, Cannon BD. 1996. *Protection Factor Testing of the Responder Suit*. Report No. ERDEC-TR-312. Aberdeen, MD: Edgewood Research, Development, and Engineering Center.

Argonne National Laboratory. 1994. *Personal Protective Equipment for the Chemical Stockpile Emergency Preparedness Program: A Status Report*. Argonne, IL: Argonne National Laboratory.

Argonne National Laboratory (ANL), Lawrence Berkeley National Laboratory (LBNL), Lawrence Livermore National Laboratory (LLNL), and Los Alamos National Laboratory (LANL), contributors. 1997. *Survey and Discussion of Models Applicable to the Transport and Fate Thrust Area of the Department of Energy Chemical and Biological Nonproliferation Program*. LANL: LA-CP-166; ANL: ANL/EAD/TM-72; LLNL: UCRL-ID-128210; and LBNL: LBNL-40764. Washington, DC: United States Department of Energy, Chemical and Biological Nonproliferation Program Office. September.

Barnea ER, Shklyar B, Moskowitz A, Barnea JD, Sheth K, Rose FV. 1995. Expression of quinone reductase activity in embryonal and adult porcine tissues. *Biological Reproduction*. 52(2):433–437.

Barnea ER, Barnea JD, Pines M. 1996. Control of cell proliferation by embryonal-origin factors. *American Journal of Reproductive Immunology*. 35(4):318–324.

Barry MA, Johnston SA. 1997. Biological features of genetic immunization. *Vaccine* 15(8):788–791.

Baskin SI, Horowitz AM, Nealley EW. 1992. The antidotal action of sodium nitrite and sodium thiosulfate against cyanide poisoning. *Journal of Clinical Pharmacology* 32(4):368–375.

Battelle Laboratories, Inc. 1993. *Final Report for Agent Testing Commercial Filters to United States Army/ERDEC.* Columbus, OH: Battelle Laboratories, Inc.

Bauer TJ, Gibbs RL. 1996. *Software User's Manual for the Chemical/Biological Agent Vapor, Liquid, and Solid Tracking (VLSTRACK) Computer Model, Version 2.1.2 (X-Window).* NSWCDD/MP-96/203. Dahlgren, VA: System Research and Technology Department, Naval Surface Warfare Center, Dahlgren Division.

Belgrader P. 1998. Autonomous Pathogen Detection System: JFT IV Field Test Results. Abstract distributed at DOE Chemical and Biological Nonproliferation Program Summer Meeting, July 28–30, McLean, VA.

Belgrader P, Smith JK, Weedn VW, Northrop MA. 1998. Rapid PCR for identity testing using a battery-powered miniature thermal cycler. *Journal of Forensic Science* 43:315–319.

Belmonte RB. 1998. *Tests of Level A Suits—Protection Against Chemical and Biological Warfare Agents and Simulants: Executive Summary.* Aberdeen Proving Ground, MD: Edgewood Research, Development and Engineering Center (SCBRD-EN).

Benenson AS (Ed). 1995. *Control of Communicable Diseases Manual.* 16th Edition. Washington, DC: American Public Health Association.

Benschop HP, van der Schans GP, Noore D, Fidder A, Mars-Groenendijk RH, de Jong LP. 1997. Verification of exposure to sulfur mustard in two casualties of the Iran–Iraq conflict. *Journal of Analytical Toxicology* 21(4):249–251.

Beregi JP, Riou B, Lecarpentier Y. 1991. Effects of hydroxocobalamin on rat cardiac papillary muscle. *Intensive Care Medicine* 17:175–177.

Berlin CM. 1976. The treatment of cyanide poisoning in children. *Pediatrics* 46:793–796.

Bhattacharya R. 1995. Therapeutic efficacy of sodium nitrite and 4-dimethylaminophenol or hydroxylamine coadministration against cyanide poisoning in rats. *Human Experimental Toxicology* 14:29–33.

Bhattacharya R, Vijayaraghavan R. 1991. Cyanide intoxication in mice through different routes and its prophylaxis by α-ketoglutarate. *Biomedical Environmental Science* 4:452–459.

Bismuth C, Baud FJ, Pontal PG. 1988. Hydroxocobalamin in chronic cyanide poisoning. *Journal of Toxicology and Clinical Experimentation* 8:35–38.

Bisson JI, Deahl MP. 1994. Psychological debriefing and prevention of post-traumatic stress. *British Journal of Psychiatry* 165:717–720.

Blachere NE, Li Z, Chandawarkar RY, Suto R, Jaikaria NS, Basu S, Udono H, Srivastava PK. 1997. Health shock protein-peptide complexes, reconstituted in vitro, elicit peptide-specific cytotoxic T lymphocyte response and tumor immunity. *Journal of Experimental Medicine* 186(8):1315–1322.

Black RM, Clarke RJ, Harrison JM, Read RW. 1997a. Biological fate of sulphur mustard: Identification of valine and histidine adducts in haemoglobin from casualties of sulphur mustard poisoning. *Xenobiotica* 27:499–512.

Black RM, Clarke RJ, Read RW, Reid MTJ. 1994. Application of gas chromatography-mass spectrometry and gas chromatography-tandem mass spectrometry to the analysis of chemical warfare samples found to contain residues of the nerve agent sarin, sulphur mustard and their degradation products. *Journal of Chromatography A* 662:301–321.

Black RM, Harrison JM, Read RW. 1997b. Biological fate of sulphur mustard: In vitro alkylation of human haemoglobin by sulphur mustard. *Xenobiotica* 27:11–32.

Boiarski AA, Bowen GW, Durnford J, Kenny DV, Shaw MJ. 1995. *State-of-the-Art Report on Biological Warfare Agent Detection Technologies* (Rep. No. SPO900-94-D-0002). Ft. Belvoir, VA: Chemical and Biological Defense Information Analysis Center.

Bonierbale E, Debordes L, Coppet L. 1997. Application of capillary gas chromatography to the study of hydrolysis of the nerve agent VX in rat plasma. *Journal of Chromatography B: Biomedical Applications* 688:255–264.

Bowler RM, Huel D, Mergler G, Cone JE. 1994. Psychological, psychosocial, and psycho-physiological sequelae in a community affected by a railroad chemical disaster. *Journal of Traumatic Stress* 7:601–624.

Bowler RM, Ngo L, Hartney C, Lloyd K, Tager I, Midtling J, Huel G. 1997. Epidemiological health study of a town exposed to chemicals. *Environmental Research* 72:93–107.

Bowler RM, Hartney C, Ngo LH. In press. Amnestic disturbance and PTSD in the aftermath of a chemical release. *Archives of Clinical Neuropsychology.*

Boyle RE, Laughlin Jr LL. 1995. History and Technical Evaluation of the U.S. Bio/Toxin Detection Program. Arlington, VA: Battelle Memorial Institute.

Breen PH, Isserles SA, Tabac E, et al. 1996. Protective effect of stroma-free methemoglobin during cyanide poisoning in dogs. *Anesthesiology* 85:558–564.

Bridgeman MME, Marsden M, MacNee W, et al. 1991. Cysteine and glutathione concentra-tions in plasma and bronchoalveolar lavage fluid after treatment with N-acetylcysteine. *Thorax* 46:39–42.

Broad WJ. 1998. Sowing death: A special report; how Japan germ terror alerted world. *New York Times*, May 26:1.

Broomfield CA, Maxwell RP, Solana RP, et al. 1991. Protection by butyrylcholinesterase against organophosphorus poisoning in nonhuman primates. *Journal of Pharmacology and Experimental Therapeutics* 259:633–638.

Brouard A, Blaisot B, Bismuth C. 1987. Hydroxocobalamin in cyanide poisoning. *Journal of Toxicology and Clinical Experimentation* 7:155–168.

Bryden WA, Fenselau C, Cotter, RJ. 1998. TinyTOF Mass Spectrometer for Biological Detec-tion. Presentation at the Defense Advanced Research Projects Agency Meeting on Bio-surveillance: Providing Detection in the New Millenium. Johns Hopkins University Applied Physics Laboratory, Laurel, MD, February 11.

Byrne WR. 1997. Q Fever. In Sidell FR, Takafuji ET, Franz DR (Eds). *Textbook of Military Medicine Part I, Warfare, Weaponry, and the Casualty: Medical Aspects of Chemical and Biological Warfare*. Washington, DC: Office of the Surgeon General, Department of the Army, United States of America. Pp. 523–538.

Candrian U. 1995. Polymerase chain reaction in food microbiology. *Journal of Microbiological Methods* 23:89–103.

Cao LK, Anderson GP, Ligler FS, Ezzell J. 1995. Detection of *Yersinia pestis* fraction 1 antigen with a fiber optic biosensor. *Journal of Clinical Microbiology* 33:336–341.

Carmeli A, Liberman N, Mevorach L. 1994. Anxiety-related somatic reactions during missle attacks. In Danon YL and Shemer J (Eds) *Chemical Warfare Medicine*. New York: Gefen. Pp. 186–190.

Carpentier P, Foquin-Tarricone A, Bodjarian N, et al. 1994. Anticonvulsant and antilethal effects of the phencyclidine derivative TCP in soman poisoning. *Neurotoxicology* 15:837–852.

Centers for Disease Control and Prevention (CDC). 1988. Management of patients with suspected Viral Hemorrhagic Fever. *Morbidity and Mortality Weekly Report* 37(Suppl. S-3):1–15.

CDC. 1993. Preliminary report: Foodborne outbreak of *Escherichia coli* O157:H7 infections from hamburgers,Western United States. *Morbidity and Mortality Weekly Report* 42:851–886.

CDC. 1994a. *Addressing Emerging Infectious Disease Threats: A Prevention Strategy for the United States*. Atlanta: United States Department of Health and Human Services.

CDC. 1994b. Hantavirus Pulmonary Syndrome, United States, 1993. *Morbidity and Mortality Weekly Report* 43:45–48.

CDC. 1995a. CDC recommendations for civilian communities near chemical weapons depots: Guidelines for medical preparedness; Notice. *Federal Register* 60:33307–33312.

CDC. 1995b. Update: Management of patients with suspected Viral Hemorrhagic Fever, United States. *Morbidity and Mortality Weekly Report* 44:475–479.

CDC. 1997. *Additional Requirements for Facilities Transferring or Receiving Select Agents*, 42 CFR Part 72/RIN 0905-AE70. Atlanta: United States Department of Health and Human Services.

CDC and the National Institutes of Health. 1993. Biosafety in Microbiological and Biomedical Laboratories. (DHHS Publication No. CDC 93-8395). Washington, DC: U.S. Government Printing Office.

Charych D, Cheng Q, Reichert A, Kuziemko G, Stroh M, Nagy JO, Spevek W, Stevens RC. 1996. A "litmus test" for molecular recognition using artificial membranes. *Chemistry and Biology* 3:113–120.

Chemical Casualty Care Office. 1995. *Medical Management of Chemical Casualties Handbook*. Second Edition. Aberdeen, MD: U.S. Army Medical Research Institute of Chemical Defense.

Chenoweth MB, Kandel A, Johnson LB, Bennett DR. 1951. Factors influencing fluoroacetate poisoning: Practice treatment with glycerol monoacetate. *Journal of Pharmacological and Experimental Therapy* 102:31–49.

Cooper DE. 1998. Upconverting phosphor-based sensors for biological agent detection. Presentation at the Defense Advanced Research Projects Agency Meeting on Bio-surveillance: Providing Detection in the New Millenium. Johns Hopkins University Applied Physics Laboratory, Laurel, MD, February 11.

Cotter RJ. 1998. Biomarker Mass Spectrometers. Presentation at the Defense Advanced Research Projects Agency Meeting on Bio-surveillance: Providing Detection in the New Millenium. Johns Hopkins University Applied Physics Laboratory, Laurel, MD, February 11.

Cottrell JE, Casthely P, Brodie JD, et al. 1978. Prevention of nitroprusside-induced cyanide toxicity with hydroxocobalamin. *New England Journal of Medicine* 298:809–811.

Cox RD. 1994. Decontamination and management of hazardous materials exposure victims in the Emergency Department. *Annals of Emergency Medicine* 23:761–770.

Crompton R, Gall D. 1980. Georgi Markov: Death in a pellet. *Medico-Legal Journal* 48:51–62.

Curran PS, Bell P, Murray A, Loughrey G, Roddy R, Rocke LG. 1990. Psychological consequences of the Enniskillen bombing. *British Journal of Psychiatry* 156:479–482.

Curry RD, Clevenger T. 1997. New approaches to countering biological terrorism with electrotechnologies: An overview. *Proceedings Report, Conference on Countering Biological Terrorism: Strategic Firepower in the Hands of Many?* Arlington, VA: Potomac Institute for Policy Studies.

Dangwal SK. 1994. A spectrophotometric method for determination of phosgene in air. *Industrial Health* 32:41–47.

Danley D. 1997. Unpublished oral remarks on the DoD Vaccine Program. Meeting of the Committee on Improving Civilian Medical Response to Chemical and Biological Terrorism Incidents, Washington, DC. July 24.

Dart RC, Stark Y, Fulton B, Koziol-McLain J, Lowenstein SR. 1996. Insufficient stocking of poisoning antidotes by hospital pharmacies. *Journal of the American Medical Association* 276:1508–1510.

Daugherty ML, Watson AP, Vo-Dinh T. 1992. Currently available permeability and break-through data characterizing chemical warfare agents and their simulants in civilian protective clothing materials. *Journal of Hazardous Materials* 30:243–267.

Daugherty PS, Chen G, Olsen MJ, Iverson BL, Georgiou G. 1998. Antibody affinity maturation using bacterial surface display. *Protein Engineering* 11(9):825–832.

Davison CD, Roman RS, Smith PK. 1961. Metabolism of bis-β-chloroethyl sulphide (sulphur mustard gas). *Biochemical Pharmacology* 7:65–74.

Deahl MP, Bisson JI. 1995. Dealing with disasters: Does psychological debriefing work? *Journal of Accidents and Emergency Medicine* 12(4):255–258.

Deahl MP, Gillham AB, Thomas J, Searle MM, Srinivasan M. 1994. Psychological sequelae following the Gulf War: Factors associated with subsequent morbidity and the effectiveness of psychological debriefing. *British Journal of Psychiatry* 165:60–65.

Defense Protective Service. 1996. *10-90 Gold NBC Response Plan*. Washington, DC: United States Department of Defense.

Defense Special Weapons Agency (DSWA). 1997. *Hazard Prediction Assessment Capability (HPAC), Version 3.0 Software* (User Manual). Alexandria, VA: Weapons Effects Division, Defense Special Weapons Agency.

Deshpande SS, Smith CD, Filbert MG. 1995. Assessment of primary neuronal culture as a model for soman-induced neurotoxicity and effectiveness of memantine as neuroprotective drug. *Archives of Toxicology* 69:384–390.

DeLucas L. 1998. Structure-Based Drug Design for Microorganism-Associated Proteins. Presentation at the Defense Advanced Research Projects Agency Meeting on Biosurveillance: Providing Detection in the New Millenium. Johns Hopkins University Applied Physics Laboratory, Laurel, MD, February 11.

Diller WF. 1980. The methenamine misunderstanding in the therapy of phosgene poisoning. *Archives of Toxicology* 46:199–206.

Doctor BP, Blick DW, Caranto G, et al. 1993. Cholinesterases as scavengers for organophosphorus compounds: Protection of primate performance against soman toxicity. *Chemistry and Biology* 87:285–293.

Dulaney MD, Brumley M, Willis JT, Hume AS. 1991. Protection against cyanide toxicity by oral alpha-ketoglutaric acid. *Veterinary and Human Toxicology* 33:571–575.

Dulay MT, Yan C, Zare RN, Rakestraw DJ. 1995. Automated capillary electrochromatography: reliability and reproducibility studies. *Journal of Chromatography A* 725:361–366.

Dunn, MA, BE Hackley, FR Sidell. 1997. Pretreatment for nerve agent exposure. In: Sidell FR, Takafuji ET, Franz DR (Eds). *Textbook of Military Medicine Part I, Warfare, Weaponry, and the Casualty: Medical Aspects of Chemical and Biological Warfare*. Washington, DC: Office of the Surgeon General, Department of the Army, United States of America. Pp. 181–196.

Edberg L, Luo M. 1997. Crystallization of Venezuelan Equine Encephalitis Virus (VEEV) Nucleocapsid Protein (abstract). Presented at Emerging Infections and Antimicrobial Resistance: Rational Approaches to Drug Design. New Orleans, May 30–June1.

Edgewood Safety Office. 1996. Military Unique Material Safety Data Sheets. Edgewood, MD: Edgewood Research, Development, and Engineering Center. Accessed October 10, 1997 at http://www.cbdcom.apgea.army.mil/RDA/erdec/risk/safety/msds/

Ellman GL, Courtney KD, Andres Jr. J, Featherstone RM. 1961. A new and rapid colorimetric determination of acetylcholinesterase activity. *Biochemical Pharmacology* 7:88–95.

Ember LR. 1984. Yellow rain. *Chemical and Engineering News* 62:8–34.

Ermak DL. 1998. Modeling and prediction thrust overview. Abstract distributed at DOE Chemical and Biological Nonproliferation Program Summer Meeting, July 28–30, McLean, VA.

Evans ME, Friedlander AM. 1997. Tularemia. In: Sidell FR, Takafuji ET, Franz DR (Eds). *Textbook of Military Medicine Part I, Warfare, Weaponry, and the Casualty: Medical Aspects of Chemical and Biological Warfare.* Washington, DC: Office of the Surgeon General, Department of the Army, United States of America. Pp. 503–512.

Fainberg A. 1997. Debating policy priorities and implications. In: Roberts B (Ed). *Terrorism With Chemical and Biological Weapons.* Alexandria, VA: The Chemical and Biological Arms Control Institute. Pp. 75–93.

Farchaus JW, Ribot WJ, Jendrek S, Little SF. 1998. Fermentation, purification, and characterization of protective antigen from a recombinant, avirulent strain of *Bacillus anthracis. Applied Environmental Microbiology* 64(3):892–891.

Federal Emergency Management Agency. 1997. *Terrorist Incident Annex to the Federal Response Plan*, Document No. FEMA 229. Washington, DC: Federal Emergency Management Agency.

Feldstein M, Klendshoj NC. 1954. The determination of cyanide in biological fluids by microdiffusion analysis. *Journal of Laboratory and Clinical Medicine* 44:166–170.

Fenselau C. 1997. MALDI MS and strategies for protein analysis. *Analytical Chemistry* 69:661A–665A.

Fenselau C. 1998. Rapid Characterization of Biomarkers by Mass Spectrometry. Presentation at the Defense Advanced Research Projects Agency Meeting on Bio-surveillance: Providing Detection in the New Millenium. Johns Hopkins University Applied Physics Laboratory, Laurel, MD, February 11.

Fidder A, Noort D, deJong AL, Trap HC, deJong LP, Benschop HP. 1996a. Monitoring of in vitro and in vivo exposure to sulfur mustard by GC/MS determination of the N-terminal valine adduct in hemoglobin after a modified Edman degradation. *Chemical Research and Toxicology* 9:788–792.

Fidder A, Noort D, deJong LP, Benschop HP, Hulst AG. 1996b. N7-(2-hydroxyethylthioethyl)-guanine: A novel urinary metabolite following exposure to sulphur mustard. *Archives of Toxicology* 10:854–855.

Flynn BW. 1996. Psychological aspects of terrorism. Presentation at the First Harvard Symposium on the Medical Consequences of Terrorism. Boston. Accessed at Website http://www.mentalhealth.org/emerserv/terroris.htm February 26, 1998.

Foch R. 1998. Micro Unmanned Vehicles. Presentation at the Defense Advanced Research Projects Agency Meeting on Bio-surveillance: Providing Detection in the New Millenium. Johns Hopkins University Applied Physics Laboratory, Laurel, MD, February 11.

Frankovich TL, Arnon SS. 1991. Clinical trial of botulism immune globulin for infant botulism. *Western Journal of Medicine* 154:103.

Franz DR, NK Jaax. 1997. Ricin Toxin. In: Sidell FR, Takafuji ET, Franz DR (Eds). *Textbook of Military Medicine Part I, Warfare, Weaponry, and the Casualty: Medical Aspects of Chemical and Biological Warfare.* Washington, DC: Office of the Surgeon General, Department of the Army, United States of America. Pp. 631–642.

Franz DR, Jahrling PB, Friedlander AM, McClain DJ, Hoover DL, Byrne WR, Pavlin JA, Christopher GW, Eitzen EM. 1997. Clinical recognition and management of patients exposed to biological warfare agents. *Journal of the American Medical Association* 278:399–411.

Friedlander AM. 1997. Anthrax. In: Sidell FR, Takafuji ET, Franz DR (Eds). *Textbook of Military Medicine Part I, Warfare, Weaponry, and the Casualty: Medical Aspects of Chemical and Biological Warfare.* Washington, DC: Office of the Surgeon General, United States Army. Pp. 467–478.

Fume Free, Inc. *Quick Mask Description.* http://.quickmask.com/descript.htm Website accessed September 25, 1997.

Gardner JP, Zhu H, Colosi PC, Kurtzman GJ, Scadden DT. 1997. Robust, but transient expression of adeno-associated virus-transduced genes during human T lymphnopoiesis. *Blood* 90(12):4854–4864.

Garner JS and the Hospital Infection Control Practices Advisory Committee, Centers for Disease Control and Prevention. 1996. Guidelines for isolation precautions in hospitals. *Infection Control and Hospital Epidemiology* 17:53–80.

Georgiou G, Iverson BL. 1998. Miniaturized Antibodies for Ultra-High-Affinity Detection. Presentation at the Defense Advanced Research Projects Agency Meeting on Bio-surveillance: Providing Detection in the New Millenium. Johns Hopkins University Applied Physics Laboratory, Laurel, MD, February 11.

Gill DM. 1982. Bacterial toxins: A table of lethal amounts. *Microbiological Reviews* 46:86–94.

Gough AR, Markus K. 1989. Hazardous materials protections in ED Practice: Laws and logistics. *Journal of Emergency Nursing* 15:477.

Grob D, Harvey AM. 1953. The effects and treatment of nerve gas poisoning. *American Journal of Medicine* 14:52–63.

Groff WA Sr., Stemler FW, Kaminskis A, Froehlich HR, Johnson RP. 1985. Plasma-free cyanide and blood total cyanide: A rapid completely automated microdistillation assay. *Clinical Toxicology* 23:133–163.

Gross NJ. 1988. Ipratropium bromide. *New England Journal of Medicine* 319:486–494.

Guatelli JC, Gingeras TR, Richman DD. 1989. Nucleic acid amplification in vitro: Detection of sequences with low copy numbers and application to diagnosis of human immunodeficiency virus type 1 infection. *Clinical Microbiology Review* 2(2):217–226.

Guilbault GG, Schmid RD. 1991. Biosensors for the determination of drug substances. *Biotechnology and Applied Biochemistry* 14(2):133–145.

Guilbault GG, Hock B, Schmid R. 1992. A piezoelectric immunobiosensor for atrazine in drinking water. *Biosensors and Bioelectronics* 7(6):411–419.

Habermann E. 1989. Clostridial neurotoxins and the central nervous system: Functional studies on isolated preparations. In: Simpson LL (Ed). *Botulinum Toxin and Tetanus Toxin*. New York: Academic Press. Pp. 153–178.

Hall AH, Rumack BH. 1987. Hydroxocobalamin/sodium thiosulfate as a cyanide antidote. *Journal of Emergency Medicine* 5:115–121.

Hamling J. 1997. *On the Efficacy of CISD*. http://www.ozemail.com.au/~jsjp/cisd2.htm Website accessed October 10, 1997.

Hance BJ, Chess C, Sandman PM. 1988. *Improving Dialogue with Communities: A Risk Communication Manual for Government*. New Brunswick, NJ: Rutgers University.

Harrell LJ, Andersen GL, Wilson KH. 1995. Genetic variability of *Bacillus anthracis* and related species. *Journal of Clinical Microbiology* 33:1847–1850.

Harris LW, Braswell LM, Fleisher JP, Cliff WJ. 1964. Metabolites of pinacolylmethylphosphonofluridate (soman) after enzymatic hydrolysis in vitro. *Biochemical Pharmacology* 13:1129.

Higgens JA, Ezzell J, Hinnebusch BJ, Shipley M, Henchal EA, Ibrahim MS. 1998. 5′ nuclease PCR assay to detect *Yersinia pestis*. *Journal of Clinical Microbiology* 36:2284–2288.

Hillman B, Bardhan KD, Bain JTB. 1974. The use of dicobalt edetate (Kelocyanor) in cyanide poisoning. *Postgraduate Medical Journal* 50:171–174.

Hiss J, Arensburg B. 1994. Suffocation from misuse of gas masks during the Gulf War. In: Danon YL, Shemer J (Eds). *Chemical Warfare Medicine*. New York: Gefen. Pp. 106–108.

Hoffmaster AR, Koehler TM. 1997. The anthrax toxin activator geneatxA is associated with CO_2 enhanced non-toxin gene expression in *Bacillus anthracis*. *Infection and Immunity* 65:3091–3099.

Hoover KD, Friedlander AM. 1997. Brucellosis. In: Sidell FR, Takafuji ET, Franz DR (Eds). *Textbook of Military Medicine Part I, Warfare, Weaponry, and the Casualty: Medical Aspects of Chemical and Biological Warfare*. Washington, DC: Office of the Surgeon General, Department of the Army, United States of America. Pp. 513–522.

Huggins JW, Robertson M, Kefauver D, Laughlin C, Knight JC, Esposito JJ. 1996. Potential Antiviral Therapeutics for Smallpox and Other Poxvirus Infections (abstract). Presented at the XI Poxvirus and Iridovirus Meeting, Toledo, Spain, May 4–9.

Huggins JW, Marteniz MJ, Zaucha GM, Jahrling PB, Smee D, Bray, M. 1998. The DNA Polymerase Inhibitor Cidofovir (HPMC, Vistide) is a Potential Antiviral Therapeutic Agent for the Treatment Of Monkeypox And Other Orthopox Virus Infections. Presented at the XII International Poxvirus Symposium, St Thomas, VI, June 6–10.

Hume AS, Mozingo JR, McIntyre B, Ho IK. 1995. Antidotal efficacy of alpha-ketoglutaric acid and sodium thiosulfate in cyanide poisoning. *Journal of Toxicology and Clinical Toxicology* 33:721–724.

Hurst C. 1998. Decontamination. Presentation at BioScience 98. Ellicott City, MD, June 2.

Ibrahim MS, Lofts RS, Jarling PB, Henchal EA, Weedn VW, Northrup MA, Belgrader P. 1998. Real-time microchip PCR for detecting single-base differences in viral and human DNA. *Analytical Chemistry* 70:2013–2017.

Institute of Medicine. 1998. *Improving Civilian Medical Response to Chemical or Biological Terrorist Incidents: Interim Report on Current Capabilities*. Washington, DC: National Academy Press.

Institute of Medicine and National Research Council. 1998. *Ensuring Safe Food from Production to Consumption*. Washington, DC: National Academy Press.

Jackson PJ, Hugh-Jones ME, Adair DM, Green G, Hill KK, Kuske CR, Grinberg LM, Abramova FA, Klein P. 1998. PCR analysis of tissue samples from the 1979 Sverdlovsk anthrax victims: The presence of multiple *Bacillus anthracis* strains in different victims. *Proceedings of the National Academy of Sciences* 95:1224–1229.

Jahrling PB. 1997. Viral Hemorrhagic Fevers. In: Sidell FR, Takafuji ET, Franz DR (Eds). 1997. *Textbook of Military Medicine Part I, Warfare, Weaponry, and the Casualty: Medical Aspects of Chemical and Biological Warfare*. Washington, DC: Office of the Surgeon General, Department of the Army, United States of America. Pp. 591–602.

Jakubowski EM, Woodard CL, Mershon NM, Dolzine TW. 1990. Quantitation of thiodiglycolin urine by electron ionization gas chromatography-mass spectrometry. *Journal of Chromatography* 528:184–190.

Janata, J. 1989. *Principles of Chemical Sensors*. New York: Plenum Press.

Joerger RD, Truby TM, Hendrickson ER, Young, RM, Ebersole, RC. 1995. Anylate detection with DNA-labeled antibodies and polymerase chain reaction. *Clinical Chemistry* 41:1371–1377.

Johnson WS, Hall AH, Rumack BH. 1989. Cyanide poisoning successfully treated without "therapeutic methemoglobin levels." *American Journal of Emergency Medicine* 7:437–440.

Johnston SA, Barry MA. 1997. Genetic to genomic vaccination. *Vaccine* 15(8):808–809.

Kage S, Nagata T, Kudo K. 1996. Determination of cyanide and thiocyanate in blood by gas chromatography and gas chromatography-mass spectrometry. *Journal of Chromatography B: Biomedical Applications* 675:27–32.

Kassa J. 1995. Comparison of efficacy of two oximes (HI-6 and obidoxime) in soman poisoning in rats. *Toxicology* 101:167–174.

Keim P, Kalif A, Schupp J, Hill K, Travis SE, Richmond K, Adair DM, Hugh-Jones M, Kuske CR, Jackson P. 1997. Molecular evolution and diversity in *Bacillus anthracis* as detected by amplified fragment length polymorphism markers. *Journal of Bacteriology* 179:818–824.

Kenardy JA, Webster RA, Lewin TJ, Vaughan JC, Hazell PL, Carter GL. 1996. Stress debriefing and patterns of recovery following a natural disaster. *Journal of Traumatic Stress* 9:37–49.

Kingery AF, Allen HE. 1995.The environmental fate of organophosphorus nerve agents: A review. *Toxicological and Environmental Chemistry* 47:155–184.

Kirk MA, Gerace R, Kulig KW. 1993. Cyanide and methemoglobin kinetics in smoke inhalation victims treated with the cyanide antidote kit. *Annals of Emergency Medicine* 22:1413–1418.

Kluwe WM, Chinn JC, Feder P, Olson C, Joiner R. 1987. Efficacy of pyridostigmine pretreatment against acute soman intoxication in a primate model. In: *Proceedings of the Sixth Medical Chemical Defense Bioscience Review*. Report AD B121516. Aberdeen Proving Ground, MD: US Army Medical Research Institute for Chemical Defense. Pp. 277–234.

Konings DA, Wyatt JR, Ecker DJ, Freier SM. 1997. Strategies for rapid deconvolution of combinational libraries: comparative evaluation using a model system. *Journal of Medical Chemistry* 40(26):4386–4395.

Koplovitz I, Stewart JR. 1994. A comparison of the efficacy of HI6 and 2-PAM against soman, tabun, sarin, and VX in the rabbit. *Toxicology Letters* 70:269–279.

Kress-Rogers E, ed. 1997. *Handbook of Biosensors and Electronic Noses: Medicine, Food and the Environment*. Boca Raton, FL: CRC Press.

Kricka LJ. 1998. Revolution on a square centimeter. *Nature Biotechnology* 16:513–514.

Kwoh DY, Davis GR, Whitfield KM, Chappelle HL, DiMichele LJ, Gingeras TR. 1989. Transcription-based amplification system and detection of amplified human immunodeficiency virus type 1 with a bead-based sandwich hybridization format. *Proceedings of the National Academy of Sciences* 86(4):1173–1177.

Lailey AF, Hill L, Lawston IW, et al. 1991. Protection by cysteine esters against chemically induced pulmonary edema. *Biochemical Pharmacology* 42:S47–S54.

Lambert RJ, Kindler BL, Schaeffer DJ. 1988. The efficacy of superactivated charcoal in treating rats exposed to a lethal oral dose of potassium cyanide. *Annals of Emergency Medicine* 17:595–598.

Langmuir AD, Andrews JM. 1952. Biological warfare defense: The Epidemic Intelligence Service of the Communicable Disease Center. *American Journal of Public Health* 42(3):235–238.

LeBlanc FN, Benson BE, Gilgad A. 1986. Severe organophosphate poisoning requires the use of an atropine drip. *Journal of Toxicology and Clinical Toxicology* 24:69–76.

LeBlanc JF, McLane KE, Parren PW, Burton DR, Ghazal P. 1998. Recognition properties of a sequence-specific DNA binding antibody. *Biochemistry* 37(17):6015–6022.

Lederberg J, Shope RE, Oaks SC, Jr. (Eds). 1992. *Emerging Infections: Microbial Threats to Health in the United States*. Washington, DC: National Academy Press.

Lenz DE, Boisseau J, Maxwell DM, Heir E. 1987. Pharmacokinetics of soman and its metabolites in rats. *Proceedings of the 6th Medical Chemical Defense Bioscience Review* (Rep. No. AD B121516). Aberdeen, MD: U.S. Army Medical Research Institute of Chemical Defense. Pg. 201.

Lenz DE, Brimfield, AA, Cook, LA, 1997. Development of immunoassays for detection of chemical warfare agents. *American Chemical Society Series* 657:77–86.

Lenz DE, Brimfield AA, Hunter KW, et al. 1984. Studies using a monclonal antibody against soman. *Fundamental and Applied Toxicology* 4:S156–S164.

Lenz DE, Yourick JJ, Dawson JS, Scott J. 1992. Monoclonal antibodies against soman: Characterization of soman stereoisomers. *Immunology Letters* 31:131–136.

Levitin HW, Siegelson HJ. 1996. Hazardous materials: Disaster medical planning and response. *Disaster Medicine* 14:327–349.

Litovitz TL, Smilkstein M, Felberg L, Klein-Schwartz W, Berlin R, Morgan JL. 1997. 1996 Annual Report of the American Association of Poison Control Centers Toxic Exposure Surveillance System. *Toxicology* 15:447–453.

Little JS, Broomfield CA, Fox-Talbot MKF et al. 1989. Partial characterization of an enzyme that hydrolyzes sarin, soman, tabun and diisopropyl phosphorofluoridate (DFP). *Biochemical Pharmacology* 38:23–29.

Lorin HG, Kulling PEJ. 1986. The Bhopal tragedy. *Journal of Emergency Medicine* 4:311–316.

Lundquist P, Rosling H, Sorbo B. 1985. Determination of cyanide in whole blood, erythrocytes, and plasma. *Clinical Chemistry* 31:591–595.

Lundy PM, Hansen AS, Hand BT, Boulet CA. 1992. Comparison of several oximes against poisoning by soman, tabun and GF. *Toxicology* 72:99–105.

Manchee RJ, Stewart WDP. 1988. The decontamination of Gruinard Island. *Chemistry in Britain* 24(7):690–691.

Mariella R Jr. 1998. Autonomous Pathogen Detection System: Overview and System Description. Abstract distributed at DOE Chemical and Biological Nonproliferation Program Summer Meeting, July 28–30, McLean, VA.

Marrs TC, Bright JE. 1987. Effect on blood and plasma cyanide levels and on methemoglobin levels of cyanide administered with or without previous pretection using PAPP. *Human Toxicology* 6:139–145.

Marshall A, Hodgson J. 1998. DNA chips: An array of possibilities. *Nature Biotechnology* 16:27–31.

Masson P, Adkins S, Gouet P, Lockridge O. 1993. Recombinant human butyrylcholinesterase G390V, the fluoride-2 variant, expressed in chinese hamster ovary cells, is a low affinity variant. *Journal of Biological Chemistry* 268(19):14329–14341.

Martin LJ, Doebler JA, Shih T, Anthony A. 1985. Protective effect of diazepam pretreatment on soman-induced brain lesion formation. *Brain Research* 325:287–289.

Masuda N, Takatsu M, Morinari H, Ozawa T. 1995. Sarin poisoning in Tokyo subway. *Lancet* 345:1446.

Maxwell DM, Brecht KM, O'Neill BL. 1987. The effect of carboxylesterase inhibition on interspecies differences in soman toxicity. *Toxicology Letters* 39:35–42.

Maxwell DM, Castro CA, De La Hoz DM, et al. 1992. Protection of rhesus monkeys against soman and prevention of performance decrement by pretreatment with acetylcholinesterase. *Toxicology and Applied Pharmacology* 115:44–49.

Mazzola CA, Addis RP, and Emergency Management Laboratory, Oak Ridge Institute for Science and Education. 1995. *Atmospheric Dispersion Modeling Resources*. Second Edition. Washington, DC: United States Department of Energy, Emergency Management Advisory Committee, Subcommittee on Consequence Assessment and Protective Actions, Office of Nonproliferation and National Security. March.

McClain DJ. 1997. Smallpox. In: Sidell FR, Takafuji ET, Franz DR (Eds). *Textbook of Military Medicine Part I, Warfare, Weaponry, and the Casualty: Medical Aspects of Chemical and Biological Warfare*. Washington, DC: Office of the Surgeon General, Department of the Army, United States of America. Pp. 539–560.

McDonough JH Jr, Jaax NK, Crowley RA, Mays MZ, Modrow HE. 1989. Atropine and/or diazepam therapy protects against soman-induced neural and cardiac pathology. *Fundamental and Applied Toxicology* 13:256–276.

McGovern TW, Friedlander AM. 1997. Plague. In Sidell FR, Takafuji ET, and Franz DR (Eds). *Textbook of Military Medicine Part I, Warfare, Weaponry, and the Casualty: Medical Aspects of Chemical and Biological Warfare*. Washington, DC: Office of the Surgeon General, Department of the Army, United States of America. Pp. 479-502.

McLane KE, Burton DR, Ghazal P. 1955. Transplantation of a 17-amino acid alpha-helical DNA-binding domain into a antibody molecule confers sequence-dependent DNA recognition. *Proceedings of the National Academy of Sciences* 92(11):5214–5218.

McLuckey SA. 1998. Advanced Ion Trap Mass Spectrometry for Detection and Identification of Chemical/Biological Threats. Abstract distributed at DOE Chemical and Biological Nonproliferation Program Summer Meeting, July 28–30, McLean, VA.

McLuckey SA, Stephenson JL Jr, Asono KG. 1998. Ion/ion proton-transfer kinetics: implications for analysis of ions derived from electrospray of protein mixtures. *Analytical Chemistry* 70:1198–1202.

Mecsas J, Raupach B, Falkow S. 1998. The Yersinia Yops inhibit invasion of Literia, Shigella and Edwardsiella but not Salmonella into epithelial cells. *Molecular Microbiology* 28(6):1269–1281.

Meselson M, Guillemin J, Hugh-Jones M, Langmuir A, Popova I, Shelokov A, Yampolskaya O. 1994. The Sverdlovsk anthrax outbreak of 1979. *Science* 266:1202–1208.

Middlebrook JL. 1991. Production and characterization of monoclonal antibodies to conotoxin GI. *10th World Congress on Animal, Plant, and Microbial Toxins.* November 3–8, 1991, Singapore.

Middlebrook JL, Franz DR. 1997. Botulinum toxins. In: Sidell FR, Takafuji ET, Franz DR (Eds). *Textbook of Military Medicine Part I, Warfare, Weaponry, and the Casualty: Medical Aspects of Chemical and Biological Warfare.* Washington, DC: Office of the Surgeon General, Department of the Army, United States of America. Pp. 655–676.

Miller J, Broad W J. 1998. Exercise finds U.S. unable to handle germ war threat. *New York Times,* 26 April, A1.

Mirzabekov A. 1998. 3D chip development. Presentation at the Defense Advanced Research Projects Agency Meeting on Bio-surveillance: Providing Detection in the New Millenium. Johns Hopkins University Applied Physics Laboratory, Laurel, MD, February 11.

Mitchell JT, Everly GS. 1996. *Critical Incident Stress Debriefing: An Operations Manual for the Prevention of Traumatic Stress Among Emergency Services and Disaster Workers.* Ellicott City, MD: Chevron Publishing.

Morita H, Yanigasawa T, Shimizu M, Hirabayashi, H, Okudera H, Nohara M, Midorikawa Y, Mimura S. 1995. Sarin poisoning in Matsumoto, Japan. *The Lancet* 346:290–293.

Nakano K. 1995. The Tokyo sarin gas attack. *Cross-Cultural Psychology Bulletin* December:12–15.

Nardin A, Sutherland WM, Hevey M, Schmaljohn A, Taylor RP. 1998. Quantitative studies of heteropolymer-mediated binding of inactivated Marburg virus to the complement receptor on primate erythrocytes. *Journal of Immunological Methods* 211(1–2):21–31.

National Association of County and City Health Officials. 1997. *NACCHO Study of Electronic Communication Capacity of Local Health Departments.* Washington, DC: National Association of City County Health Officials.

National Research Council. 1989. *Improving Risk Communication.* Washington, DC: National Academy Press.

National Research Council. 1998. *Review of Acute Human-Toxicity Estimates for Selected Chemical-Warfare Agents.* Washington, DC: National Academy Press.

Noll GG, Hildebrand MS. 1994. *Hazardous Materials: Managing the Incident.* Stillwater, Oklahoma: International Fire Service Training Association (ISTA).

Noort D, Hulst AG, Trap HC, deJong LPA, Benschop HP. 1997. Synthesis and mass spectrometric identification of the major amino acid adducts formed between sulphur mustard and haemoglobin in human blood. *Archives of Toxicology* 71:171–178.

Norris JC, Utley WA, Hume AS. 1990. Mechanism of antagonizing cyanide-induced lethality by α-ketoglutaric acid. *Toxicology* 62:275–283.

Northrup MA, Benett B, Hadley D, Landre P, Lehew S, Richards J, Stratton P. 1998. A miniature analytical instrument for nucleic acids based on micromachined silicon reaction chambers. *Analytical Chemistry* 70:918–922.

Novales-Li P, Priddle JD. 1995. Production and characterization of separate monoclonalantibodies to human brain and erythrocyte acetylcholinesterases. *Hybridoma* 14:67–73.

Obu S. 1996. Japanese medical team briefing. In: United States Public Health Service. *Proceedings of the Seminar on Responding to the Consequences of Chemical and Biological Terrorism*, 2-21–2-26. Washington, DC: U.S. Government Printing Office.

Odoul M, Fouillet B, Nouri B, Chambon R, Chambon P. 1994. Specific determination of cyanide in blood by headspace gas chromatography. *Journal of Analytical Toxicology* 18:205–207.

Oehler GC. 1996.The growing chemical and biological weapons threat. Testimony before the Permanent Subcommittee on Investigations of the Senate Committee on Government Affairs, March 20. Accessed at Website http://www.kimsoft.com/korea/ciachem1.htm August 18, 1998.

Okumura T, Suzuki K, Fukuda A, Kohama A, Takasu N, Ishimatsu S, Hinohara S. 1998a. The Tokyo subway sarin attack: Disaster management, part 1: Community emergency response. *Academic Emergency Medicine* 5(6):613–617.

Okumura T, Suzuki K, Fukuda A, Kohama A, Takasu N, Ishimatsu S, Hinohara S. 1998b. The Tokyo subway sarin attack: Disaster management, part 2: Hospital response. *Academic Emergency Medicine* 5(6):618–624.

Okumura T, Suzuki K, Fukuda A, Kohama A, Takasu N, Ishimatsu S, Hinohara S. 1998c. The Tokyo subway sarin attack: Disaster management, part 3: National and international responses. *Academic Emergency Medicine* 5(6):625–628.

Olson KB. 1996. Overview: Recent events and responder implications. In: United States Public Health Service. *Proceedings of the Seminar on Responding to the Consequences of Chemical and Biological Terrorism*, 2-36–2-93. Washington, DC: U.S. Government Printing Office.

Omara F, Sisodia CS. 1990. Evaluation of potential antidotes for sodium fluoroacetate in mice. *Veterinary and Human Toxicology* 32:427–431.

Osterholm MT, Birkhead S, Meriwether RA. 1996. Impediments to public health surveillance in the 1990s: The lack of resources and the need for priorities. *Journal of Public Health Management and Practice* 22: 11–15.

Palleschi G, Lavagnini MG, Moscone D, Pilloton R, D'Ottavio D, Evangelisti ME. 1990. Determination of serum cholinesterase activity and dibucaine numbers by an amperometric choline sensor. *Biosensors and Bioelectronics* 5:27–35.

Parneix-Spake A, Theisen A, Roujeau JC, et al. 1993. Severe cutaneous reactions to self defense sprays. *Archives of Dermatology* 129:913.

Peterson JC, Cohen SD. 1985. Antagonism of cyanide poisoning by chlorpromazine and sodium thiosulfate. *Toxicology and Applied Pharmacology* 81:265–273.

Polhuijs M, Langenberg JP, Benschop HP., 1997. New method for retrospective detection of exposure to organophosphorous anticholinesterases: Application to alleged sarin victims of Japanese terrorists. *Toxicology and Applied Pharmacology* 146(1):156–161.

Policastro AJ. 1998. Understanding chemical/biological dispersion in subways. Abstract distributed at DOE Chemical and Biological Nonproliferation Program Summer Meeting, July 28–30, McLean, VA.

Pomerantsev AP, Staritsin NA, Mockov YV, Marinin LI. 1997. Expression of cereolysine ab genes in *Bacillus anthracis* vaccine strain ensures protection against experimental hemolytic anthrax infection. *Vaccine* 15(17/18):1848–1850.

Powers L, Ellis W Jr. 1998. Pathogenic microbe sensor technology. Presentation at the Defense Advanced Research Projects Agency Meeting on Bio-surveillance: Providing Detection in the New Millenium. Johns Hopkins University Applied Physics Laboratory, Laurel, MD, February 11.

Ramsay G. 1998. DNA chips: State of the art. *Nature Biotechnology* 16:40–44.

Ramsey JM, Jacobson SC, Knapp MR. 1995. Microfabricated chemical measurement systems. *Nature Medicine* 1(10):1093–1095.

Raphael B, Meldrum L, McFarlane AC. 1995. Does debriefing after psychological trauma work? *British Medical Journal* 310:1479–1480.

Rauber A, Heard J. 1985. Castor bean toxicity reexamined: A new perspective. *Veterinary and Human Toxicology* 27:498–502.

Raveh L, Grunwald J, Marcus D, et al. 1993. Human butyrlcholinesterase as a general prophylactic antidote for nerve agent toxicity. *Biochemical Pharmacology* 45:2465–2474.

Ray R, Boucher LJ, Broomfield CA, Lenz DE. 1988. Specific soman-hydrolyzing enzyme activity in a clonal neuronal cell culture. *Biochimica Biophysica Acta* 967:373–381.

Reynolds ML, Little PJ, Thomas BF, Bagley RB, Martin BR. 1985. Relationship between the biodisposition of (3h) soman and its pharmacological effects in mice. *Toxicology and Applied Pharmacology* 80:409.

Risatti JB, Capman WC, Stahl D. 1994. Community structure of a microbial mat: The phylogenetic dimension. *Proceedings of the National Academy of Sciences* 91:10173–10177.

Risk M, Fuortes L. 1991. Chronic arsenicalism suspected from arsine exposure: A case report and literature review. *Veterinary and Human Toxicology* 33:590.

Ro YS, Lee CW. 1991. Tear gas dermatitis. Allergic contact sensitization due to CS. *International Journal of Dermatology* 30:576–577.

Roberts JJ, Warwick GP. 1963. Studies of the mode of action of alkylating agents—VI. The metabolism of bis-2-chloroethyl sulphide (mustard gas) and related compounds. *Biochemical Pharmacology* 12:1329–1334.

Rockwood GA, Romano JA Jr., Scharf BA, Baskin SI. 1992. The effects of P-aminopropiophenone (PAPP) and P-aminooctoylphenone (PAOP) against sodium cyanide (CN) challenge on righting and motor activity in mice. *Toxicologist* 12:271.

Rogers KR, Mulchandani A, Zhou W, eds. 1995. *Biosensor and Chemical Sensor Technology: Process Monitoring and Control.* ACS Symposium Series 613. Washington, DC: American Chemical Society.

Sackett DE. 1996. Conflict simulations: Saving time, money, and lives. In: Upadhye R (Ed). *Science and Technology Review.* UCRL-52000-96-11. Livermore, CA: Lawrence Livermore National Laboratory. November. Pp. 4–11.

Saslaw S, Eigelbach HT, Prior JA, Wilson HE, Carhart S. 1961a. Tularemia vaccine study, I: Intracutaneous challenge. *Archives of Internal Medicine* 107:134–146.

Saslaw S, Eigelbach HT, Wilson HE, Prior JA, Carhart S. 1961b. Tularemia vaccine study, II: Respiratory challenge. *Archives of Internal Medicine* 107:121–133.

Sawyer TW, Lundy PM, Weiss MT. 1996. Protective effect of an inhibitor of nitric oxide synthase on sulphur mustard toxicity in *vitro*. *Toxicology and Applied Pharmacology* 141:138–144.

Scadden DT. 1997. Cytokine use in the management of HIV disease. *Journal of Acquired Immune Deficiency Syndrome Human Retrovirol* 16 Suppl 1:S23–S29.

Schneider NR. 1987. The Nebraska Veterinary Medical Association antidote depot. *Journal of the American Veterinary Medical Association* 190:797–799.

Sciuto AM, Strickland PT, Kennedy TP, et al. 1995. Protective effects of N-acetylcysteine treatment after phosgene exposure in rabbits. *American Journal of Respiratory and Critical Care Medicine* 151:768–772.

Sciuto AM, Strickland PT, Kennedy TP, Gurtner GH. 1997. Postexposure treatment with aminophylline protects against phosgene-induced acute lung injury. *Experimental Lung Research* 23:317–332.

Sextro RG. 1998. Modeling CB dispersion in buildings. Abstract distributed at DOE Chemical and Biological Nonproliferation Program Summer Meeting, July 28–30, McLean, VA.

Shalev AY. 1992. Posttraumatic stress disorder among injured survivors of a terrorist attack: Predictive value of early intrusion and avoidance symptoms. *The Journal of Nervous and Mental Disease* 180(8):505–509.

Shapira Y, Bar Y, Berkenstadt H, Atsmon J, Danon YL. 1994. Outline of hospital organization for a chemical warfare attack. In: Danon YL, Shemer J (Eds). *Chemical Warfare Medicine*. New York: Gefen. Pp. 144–151.

Shapiro RL, Hatheway C, Becher J, Swerdlow DL. 1997. Botulism surveillance and emergency response: A public health strategy for a global challenge. *Journal of the American Medical Association* 278:433–435.

Sidell FR. 1974. Soman and Sarin: Clinical manifestations and treatment of accidental poisoning by organophosphates. *Clinical Toxicology* 7:1–17.

Sidell FR. 1996. United States medical team briefing. In: United States Public Health Service. *Proceedings of the Seminar on Responding to the Consequences of Chemical and Biological Terrorism*, 2-30–2-35. Washington, DC: U.S. Government Printing Office.

Sidell FR, Markis JE, Graff WA, Kaminskis A. 1974. Enhancement of drug absorption after administration by an automatic injector. *Journal of Pharmacokinetics and Biopharmacology* 2:197–210.

Sidell FR, Takafuji ET, Franz DR (Eds). 1997. *Textbook of Military Medicine Part I, Warfare, Weaponry, and the Casualty: Medical Aspects of Chemical and Biological Warfare*. Washington, DC: Office of the Surgeon General, Department of the Army, United States of America.

Siegel LS, Johnson-Winegar AD, Sellin LC. 1986. Effect of 3,4 diaminopyridine on the survival of mice injected with botulinum neurotoxin type A, B, E or F. *Toxicology and Applied Pharmacology* 84:255–263.

Skolfield S, Lambert D, Tomassoni A, Wallace K. 1997. Inadequate regional antidote and medication supplies for poisoning emergencies. *Clinical Toxicology* 35:490.

Smith JF, Davis K, Hart MK, Ludwig GV, McClain DL, Parker MD, Pratt WD. 1997. Viral Encephalitides. In: Sidell FR, Takafuji ET, Franz DR (Eds). *Textbook of Military Medicine Part I, Warfare, Weaponry, and the Casualty: Medical Aspects of Chemical and Biological Warfare*. Washington, DC: Office of the Surgeon General, Department of the Army, United States of America. Pp. 561–590.

Sparenborg S, Brennecke LH, Jaax NK, Braitman DJ. 1992. Dizocilpine (MK-801) arrests status epilepticus and prevents brain damage induced by soman. *Neuropharmacology* 31:357–368.

Stahl D. 1998. RNA identification of bacteria. Presentation at the Defense Advanced Research Projects Agency Meeting on Bio-surveillance: Providing Detection in the New Millenium. Johns Hopkins University Applied Physics Laboratory, Laurel, MD, February 11.

Steffen R, Melling J, Woodall JP, Rollin PE, Lang RH, Luthy R, Wakdvogel A. 1997. Preparation for emergency relief after biological warfare. *Journal of Infection* 34:127–132.

Steinhaus RK, Baskin SI, Clark JH, Kirby SD. 1990. Formation of methemoglobin and metmyoglobin using 8-aminoquinoline derivatives or sodium nitrite and subsequent reaction with cyanide. *Journal of Applied Toxicology* 10:345–351.

Stern PC, Fineberg HV (Eds). 1996. *Understanding Risk: Informing Decisions in a Democratic Society*. Washington, DC: National Academy Press.

Stephenson JL Jr, McLuckey SA. 1998. Charge manipulation for improved mass determination of high-mass species and mixture components by electrospray mass spectrometry. *Journal of Mass Spectrometry* 33:664–672.

Stiles BG. 1993. Acetylcholine receptor binding characteristics of snake and cone snail venom postsynaptic neurotoxins: Further studies with a non-radioactive assay. *Toxicon* 31:825–834.

Sullivan J, Krieger G. 1992. *Hazardous Materials Toxicology*. Baltimore: Williams and Wilkins.

Taylor RF, Schultz JS, eds. 1996. *Handbook of Chemical and Biological Sensors*. Philadelphia: Institute of Physics Publications.

Taylor RP, Sutherland WM, Martin EN, Ferguson PJ, Reinagel, ML, Gilbert E, Lopez K, Incardona NL, Ochs HD. 1997. Bispecific monoclonal antibody complexes bound to primate erythrocyte complement receptor 1 facilitate virus clearance in a monkey model. *The Journal of Immunology* 158:842–850.

Ten Eyck RP, Schaerdel AD, Lynett JE, et al. 1983. Stromafree methemoglobin solution as an antidote for cyanide poisoning: A preliminary study. *Journal of Toxicology and Clinical Toxicology* 21:343–358.

Ten Eyck RP, Schaerdel AD, Ottinger WE. 1986. Comparison of nitrate treatment and stroma-free methemoglobin solution as antidotes for cyanide poisoning in a rat model. *Journal of Toxicology and Clinical Toxicology* 23:477–487.

Tuerk C, MacDougal-Waugh S. 1993. In vitro evolution of functional nucleic acids: high-affinity RNA ligands of HIV-1 proteins. *Gene* 137:33–39.

Turnbough C, Kearney J. 1998. Capture and Detection of Bacillus Spores. Presentation at the Defense Advanced Research Projects Agency Meeting on Bio-surveillance: Providing Detection in the New Millenium. Johns Hopkins University Applied Physics Laboratory, Laurel, MD, February 11.

Turner PF, Karube I, Wilson GS. 1987. *Biosensors: Fundamentals and Applications*. New York: Oxford University Press.

Ulrich RG, Sidell S, Taylor TJ, Wilhelmsen CL, Franz DR. 1997. Staphylococcal Enterotoxin B and related Pyrogenic toxins. In: Sidell FR, Takafuji ET, Franz DR (Eds). *Textbook of Military Medicine Part I, Warfare, Weaponry, and the Casualty: Medical Aspects of Chemical and Biological Warfare*. Washington, DC: Office of the Surgeon General, Department of the Army, United States of America. Pp. 621–630.

United States Army. 1995. *Medical Management of Chemical Casualties Handbook* (2nd. ed.). Aberdeen Proving Ground, MD: Chemical Casualty Care Office, U.S. Army.

United States Army Chemical Demilitarization and Remediation Activity. 1994. *Personal Protective Equipment (PPE) Alternatives for Non-Stockpile Operations Test Report*. Aberdeen, MD: United States Army Chemical Demilitarization and Remediation Activity.

United States Department of Health and Human Services (U.S. DHHS). 1993. *Hazardous Substances Emergency Events Surveillance (HSEES) Annual Report*. Atlanta: Agency for Toxic Substances and Disease Registry.

U.S. DHHS. 1994a. *Medical Management Guidelines for Acute Chemical Exposures*. Atlanta: Agency for Toxic Substances and Disease Registry.

U.S. DHHS. 1994b. *Hazardous Substances Emergency Events Surveillance (HSEES) Annual Report 1994*. Atlanta: Agency for Toxic Substances and Disease Registry.

U.S. DHHS. 1995a. *Hazardous Substances Emergency Events Surveillance (HSEES) Annual Report 1995*. Atlanta: Agency for Toxic Substances and Disease Registry.

U.S. DHHS. 1995b. Recommendations for Civilian Communities near Chemical Weapons Depots: Guidelines for Medical Preparedness. *Federal Register* 60(June 27):33308–33312.

United States Environmental Protection Agency and National Oceanic and Atmospheric Administration. 1996. *ALOHA (Areal Locations of Hazardous Atmospheres), MARPLOT 3.1, and CAMEO (Computer-Aided Management of Emergency Operations) User's Manuals.* Chemical Preparedness and Prevention Office, United States Environmental Protection Agency, Washington, DC, and Hazardous Materials Response and Assessment Division, National Oceanic and Atmospheric Administration, Seattle, WA. March.

United States Food and Drug Administration. 1997. Accessibility to new drugs for use in military and civilian exigencies when traditional human efficacy studies are not feasible. *Federal Register* 62:40996–41001.

Vale JA, Meredith TJ, Heath A. 1990. High doses of atropine in organophosphorus poisoning. *Postgraduate Medical Journal* 66:881.

Van Emon JM, Gerlach CL, Johnson JC. 1996. *Environmental Immunochemical Models: Perspectives and Applications.* ACS Symposium Series 646. Washington, DC: American Chemical Society.

Velan B, Kronman C, Grosfeld H, et al. 1991. Recombinant human acetylcholinesterase is secreted from transiently transfected 293 cells as a souble globular enzyme. *Cellular and Molecular Neurobiology* 11:143–155.

Vitko J Jr, Kottenstette R. 1998. Parallel Micro Separations-based Detection of Biotoxins and Chemical Agents. Abstract distributed at DOE Chemical and Biological Nonproliferation Program Summer Meeting, July 28–30, McLean, VA.

Vojvodic V, Milosavljevi Z, Boskovic B, Bojani CN. 1985. The protective effect of different drugs in rats poisoned by sulfur and nitrogen mustards. *Fundamentals of Applied Toxicology* 5:S160–S168.

Vyner HM. 1988. The psychological dimensions of health care for patients exposed to radiation and the other invisible environmental contaminants. *Social Science and Medicine* 27:1097–1103.

Waelbroeck M, Tasknoy M, Camus J, Christophe J. 1991. Binding kinetics of quinuclidinyl benzilate and methyl quinuclidinyl benzilate evantiomes at neuronal (M1), cardiac (M2), and pancreatic (M3) muscarinic receptors. *Molecular Pharmacology* 40:413–420.

Walker GT, Fraiser MS, Schram JL, Little MC, Nadeau JG, Malinowski DP. 1992. Strand displacement amplification—an isothermal, in vitro DNA amplification technique. *Nucleic Acids Research* 20(7):1691–1696.

Wannamacher RW Jr, Bunner DL, Neufeld HA. 1991. Toxicity of Trichothecenes and other related Mycotoxins in laboratory animals. In: Smith JE, Henderson RS (Eds). *Mycotoxins and Animal Foods.* Boca Raton, FL: CRC Press. Pp. 499–552.

Wannamacher RW Jr, Weiner SL. 1997. Trichothecene Mycotoxins. In: Sidell FR, Takafuji ET, Franz DR (Eds). *Textbook of Military Medicine Part I, Warfare, Weaponry, and the Casualty: Medical Aspects of Chemical and Biological Warfare.* Washington, DC: Office of the Surgeon General, Department of the Army, United States of America. Pp. 655–676.

Weger NP. 1983. Treatment of cyanide poisoning with 4-DMAP—Experimental and clinical overview. *Fundamental and Applied Toxicology* 3:387–396.

Weiner SL. 1996. Strategies for the prevention of a successful biological warfare aerosol attack. *Military Medicine* 161:251–256.

Weisaeth L. 1989. Importance of high response rates in traumatic stress research. *Acta Psychiatrica Scandinavia Supplementum* 355:131–137.

WHO Group of Consultants. 1970. *Health Aspects of Chemical and Biological Weapons.* Geneva: World Health Organization.

Wick CH, Yeh HR, Carlon HR, Anderson D. 1997. *Virus Detection: Limits and Strategies* (Tech. Rep. ERDEC-TR-453). Aberdeen MD: Edgewood Research Development and Engineering Center.

Wiley J, Balmier D, Farina P, et al. 1995. Severe pulmonary injury in an infant after pepper gas self defense exposure. *Journal of Toxicology and Clinical Toxicology* 33:519. (Abstract)

Willems J. 1991. Clinical management of Mustard Gas casualties. *Annales Medicinae Belgicae* 3(Suppl.):1–61.

Wittwer CT, Ririe KM, Andrew RV, David DA, Gundry RA, Balis UJ. 1997. The Light-Cycler(tm): A microvolume multisample fluorimeter with rapid temperature control. *BioTechniques* 2:176–181.

Wolfbeis OS. 1991. *Fiber Optic Chemical Sensors and Biosensors.* Vols. I and II. Boca Raton, FL: CRC Press.

Wollenberger LV, Yao YM, Mufti NA, Schneider LV. 1997. Detection of DNA Using Upconverting Phosphor Reporter Probes. In Cohn GE and Soper, SA (Eds), *SPIE Ultrasensitive Biochemical Diagnostics II.* Bellingham, WA: Society for Photo-Optical Instrumentation and Engineering. Pp. 100–111.

Worek F, Kirchner T, Szinicz L. 1995. Effect of atropine and bispyridinium oximes on respiratory and circulatory function in guinea-pigs poisoned by sarin. *Toxicology* 95:123–133.

Wright, WH, Mufti NA, Tagg NT, Webb RR, Schneider LV. 1997. High-sensitivity Immunoassay Using a Novel Upconverting Phosphor Reporter. In Cohn GE and Soper, SA (Eds), *SPIE Ultrasensitive Biochemical Diagnostics II.* Bellingham, WA: Society for Photo-Optical Instrumentation and Engineering. Pp. 248–255

Yanagisawa N. 1996. Matsumoto, Japan (June 1994). In: United States Public Health Service. *Proceedings of the Seminar on Responding to the Consequences of Chemical and Biological Terrorism,* 2-12–2-20. Washington, DC: U.S. Government Printing Office.

Yershov G, Barsky V, Belgovskiy A, Kirillov E, Kreindlin E, Ivanov I, Parinov S, Guschin D, Drobishev A, Dubiley S, Mirzabekov, A. 1996. DNA analysis and diagnostics on oligonucleotide microchips. *Proceedings of the National Academy of Sciences* 93:4913–4918.

Zelikoff A. 1998. Technology Development in Physical Protection/Decontamination. Oral presentation to the National Research Council Meeting on Strategies to Protect the Health of U.S. Forces. Washington, D.C., April 29.

APPENDIXES

APPENDIX
A

Committee and Staff Biographies

COMMITTEE BIOGRAPHIES

Peter Rosen, M.D., FACS, FACEP (*Chair*), is Director of Emergency Medicine Residency Program at the University of California, San Diego. Dr. Rosen previously served as chair of the IOM Committee on Treatment of Near-Drowning Victims. He has authored or edited a dozen textbooks on aspects of emergency medicine and since 1983 has served as Editor-in-Chief of the *Journal of Emergency Medicine*. He is a Fellow of the American College of Surgeons, the American College of Emergency Physicians, and the American Burn Association. Dr. Rosen is an IOM member.

Leo G. Abood, Ph.D., was Professor of Pharmacology, Department of Pharmacology and Physiology, University of Rochester Medical Center until his death in January 1998. Dr. Abood was an expert on the biochemistry and physiology of the nervous system whose research focused on the isolation and characterization of neurotransmitter receptors from the mammalian brain, specifically nicotine, vasopressin, and opioid receptors. He previously served on the NRC Committee on Toxicology's Panel on Anticholinergic Compounds and the Chemical Weapons Stockpile Assessment Panel.

Georges C. Benjamin, M.D., FACP, is Deputy Secretary for Public Health Services for the State of Maryland. Dr. Benjamin was formally Commissioner of Health for the District of Columbia and, a former Chairman,

Ambulatory Care, D.C. General Hospital. From 1983 to 1987 he was Chief of Emergency Medicine at the Walter Reed Army Medical Center. He is a fellow of the American College of Physicians.

Rosemarie Bowler, Ph.D., is Assistant Professor and Fieldwork Coordinator, Department of Psychology, San Francisco State University. Dr. Bowler has done extensive research on individual and community reactions to toxic chemical spills and has chaired a recent symposium on the topic for the Agency for Toxic Substances and Disease Registry. Her clinical experience at SFSU includes assessing patients and groups of workers exposed to neurotoxins.

Jeffrey I. Daniels, D.Env., is Risk Sciences Group Leader, Health and Ecological Assessment Division, Earth and Environmental Sciences Directorate, Lawrence Livermore National Laboratory. His expertise is risk assessment and his research involves the potential human health risks from contaminated environmental media, including air, water, soil, vegetation, and the development of a coupled chemical/biological system to degrade high explosives in demilitarization waste water. He is Past-President of the Northern California Chapter of the Society for Risk Analysis.

Craig A. DeAtley, B.S., P.A., is Director of the Emergency Medical Services Program, Associate Professor, Department of Emergency Medicine and Health Care Sciences Program, and CoDirector for Hazmat Medical Services, George Washington University. He is also Deputy Medical Director, Flight Medic and SWAT Medic, Fairfax County Police; Medical Specialist, Metropolitan Medical Strike Team DC-1 (PHS-sponsored NBC responders in Washington, D.C., area); and EMS Captain, Fairfax Fire and Rescue. He serves on the editorial boards of *Rescue EMS News* and *Prehospital and Disaster Medicine*.

Lewis Goldfrank, M.D., FACMT, FACP, FACEP, is Director of Emergency Medicine, New York University School of Medicine, Bellevue Medical Center. He is the medical director of the New York City Poison Control Center. Dr. Goldfrank served as president of the Society of Academic Emergency Medicine and chairs the American Board of Emergency Medicine's Subboard on Medical Toxicology. He is coeditor of the Agency for Toxic Substances Disease Registry's *Medical Guidelines for Managing Hazmat Incidents*, and senior editor of *Goldfrank's Toxicologic Emergencies*, a standard text in medical toxicology. Dr. Goldfrank is an IOM member.

Jerome M. Hauer, M.H.S., is Director, Office of Emergency Management, City of New York. He previously was Director of Emergency Medical

Services and Emergency Management for the State of Indiana. He also directed Hazmat response, crisis management, and fire safety for IBM. He is a former Army Captain assigned to the Walter Reed Army Institute of Research and past Chair of the U.S. Earthquake Consortium. Hauer also served on the U.S. Geological Survey ad hoc working group on earthquake-related casualties. He currently serves on the FBI Scientific Advisory Council on Hazardous Materials Response.

Karen Larson, Ph.D., is a Toxicologist, Office of Toxic Substances, Washington (State) Department of Health. A molecular biologist, Dr. Larson is the Washington Health Department's liaison with the state emergency planning agency, advising on methods of detection, protection, and treatment in real and hypothetical chemical or biological disasters.

Matthew S. Meselson, Ph.D., is Thomas Dudley Cabot Professor of the Natural Sciences, Department of Molecular and Cellular Biology, Harvard University. Dr. Meselson is a member of the NAS Committee on International Security and Arms Control and the Working Group on Biological Weapons Control. He served on the NAE Committee on Alternative Chemical Demilitarization Technologies and the Advisory Panel on the Chemical Research, Development and Engineering Center. Dr. Meselson is a member of both IOM and NAS.

David H. Moore, D.V.M., Ph.D., is Director, Medical Toxicology Programs for Battelle Memorial Institute's National Security Division since January, 1998. This position follows a distinguished career of more than 20 years as a scientist in Army medical research and development, culminating in his service as Deputy Director of the U.S. Army Medical Research Institute of Chemical Defense. Dr. Moore also served as the Army Surgeon General's Advisor on Toxicology and Consultant on Comparative Medicine. Dr. Moore graduated with honors from the University of Georgia College of veterinary medicine in 1977, and earned his Ph.D. in Physiology at Emory University in 1984.

Dennis M. Perrotta, Ph.D., is Chief, Bureau of Epidemiology, Texas Department of Health. Dr. Perrotta administers the Texas Poison Center Network, serves on the Armed Forces Epidemiology Board (AFEB), and recently prepared a report for the AFEB on mustard gas and sarin. In addition, he has served on review sections at NIH and ATSDR and served as a reviewer for the IOM report on Emerging Infectious Diseases.

Linda Powers, Ph.D., is Director, National Center for the Design of Molecular Function, Professor of Electrical and Biological Engineering, and

Adjunct Professor of Physics at Utah State University. After completing her M.A. in Physics and Ph.D. in Biophysics at Harvard University, she became a member of the technical staff at AT&T Bell Laboratories. She joined the USU faculty in 1988. She has a broad scope of expertise from biochemistry to electrical engineering, and has considerable experience in heme protein catalysis, structural biology, and the design and construction of optical and X-ray instrumentation. She was a pioneer the use of X-ray absorption spectroscopy for the investigation of biological problems and has authored more than 100 technical publications in refereed journals and books.

Philip K. Russell, M.D., is Professor of International Health, School of Hygiene and Public Health, Johns Hopkins University. He is former Commander of Army Medical Research and Development. An infectious disease specialist with particular expertise in vaccines, he serves on the Scientific Advisory Board of the National Center on Infectious Disease at the Centers for Disease Control and Prevention.

Jerome Schultz, Ph.D., is Director, Center for Biotechnology and Bioengineering, University of Pittsburgh. Dr. Schultz is a biochemical engineer with expertise in biochemistry. His research is focused on using bio-molecules with recognition capability for biosensor probe devices. He is a past president of the American Institute for Medical and Biological Engineering, and is currently Vice-Chair, Board on Army Science and Technology's (BAST) Committee for the Review of Army Chemical and Biological Defense Command. Dr. Schultz is an NAE member.

Robert E. Shope, M.D., is Professor of Pathology in the WHO Center for Tropical Diseases at the University of Texas Medical Branch at Galveston. The Center serves as the repository for a major collection of arboviruses and rodent-associated viruses. He is a virologist/epidemiologist and former Director of the Yale Arbovirus Research Unit. He was involved as a member of the teams that investigated outbreaks of Rift Valley fever, Lassa fever, Venezuelan hemorrhagic fever, and other often fatal hemorrhagic diseases caused by viruses that have bioterrorism potential. He also has expertise in diagnosis and rapid identification of human-pathogenic viruses carried by arthropods and rodents, and he cochaired in 1992 the Institute of Medicine's study on emerging infections.

Robert S. Tharratt, M.D., FACP, FCCP, FACMT, is Associate Professor of Medicine and Chief, Section of Clinical Pharmacology and Medical Toxicology, Division of Pulmonary and Critical Care Medicine, University of California Davis. Dr. Tharratt is also Associate Regional Medical Director of the Davis Division of the California Poison Control System, Medical Director of Sacramento County Emergency Medical Services, and Medical Director of the Sacramento City and County Fire Agencies. He is a hazardous materials specialist and a Medical Manager for FEMA Urban Search and Rescue Team CA-7.

STAFF BIOGRAPHIES

Frederick J. Manning, Ph.D., is a Senior Program Officer in IOM's Health Science Policy Program and Study Director. In 5 years at IOM, he has served as Study Director for projects addressing a variety of topics from medical isotopes to potential hepatitis drugs, blood safety and availability, rheumatic disease, and resource sharing in biomedical research. Prior to joining IOM, Dr. Manning spent 25 years in the U.S. Army Medical Research and Development Command, serving in positions that included Director of Neuropsychiatry at the Walter Reed Army Institute of Research and Chief Research Psychologist for the Army Medical Department. Dr. Manning earned his Ph.D. in Psychology from Harvard University in 1970, following undergraduate education at the College of the Holy Cross.

C. Elaine Lawson, M.S., is a Program Officer in the Institute of Medicine's Health Sciences Section. Ms. Lawson obtained her B.S. in Physical and Health Education from James Madison University, and her M.S. in Exercise Science and Health from George Mason University. She has written chapters for IOM reports on genetic risks and stalking behavior, and was coeditor of IOM reports on healthcare in schools and on gender differences in susceptibility to environmental factors. She recently codirected the Institute of Medicine and Smithsonian Institution joint venture for a leadership institute on K–6 science education.

Carol A. Maczka, Ph.D., is the Director of Toxicology and Risk Assessment in the NAS Board on Environmental Studies and Toxicology (BEST). She obtained her Ph.D. in pharmacology from the George Washington University, with a minor concentration in the metabolism of xenobiotics. She has written a chapter for the IOM report *Veterans and Agent Orange* and participated in numerous BEST studies. Other current projects involve: drinking water contaminants, hormonally active agents in the environment, developmental toxicology, and strategies to protect the health of deployed U.S. forces. Prior to joining the NRC, Dr. Maczka was Senior Vice President of Clement International, a health and environmental consulting firm.

Inventory of Chemical and Biological Defense Technology, with Gap and Overlap Analysis

Personal Protective Equipment

07-Oct-98

Type	Product	Location/PI
Breathing		
	RP51A Respirator canister	Cabot Safety Products
	PBE (Protective Breathing Equipment)	Essex PB&R Corp.
	SCU (Self-Contained Unit)	Essex PB&R Corp.
	VRU (Victim Rescue Unit)	Essex PB&R Corp.
	Plus 10 Filter Breathing Unit	Essex PB&R Corp.
	Escape hood/mask for VIPs	Fume Free, Inc
	QuickMask Respiratory Protective Escape Device	Fume-Free, Inc.
	FRENZY AIR 5000 breathing apparatus	Giat Industries (France)
	Respiratory protection filter kits	Giat Industries
	SPIROMATIC 90	Giat Industries
	Recirculation Filter Blower	ILC Dover, Inc.
	CAPS (Civilian Adult Protective System)	Israel Ministry of Defense Export Organization (SIBAT)
	CHIPS (Chemical Infant Protective System)	Israel Ministry of Defense Export Organization (SIBAT)
	Children Hood Blower System	Israel Ministry of Defense Export Organization (SIBAT)
	Advanced Crew Member Blower System	Israel Ministry of Defense Export Organization (SIBAT)
	Portable Blower Infant Protective Crib	Israel Ministry of Defense Export Organization (SIBAT)
	M17 series masks	MSA Defense Products
	Respirator canister Model 800375	MSA Safety Products

Type	Product	Location/PI
Breathing		
	ESP Mask Communication System	MSA Safety Products
	Escort (SCBA) Escape Self Contained Breathing Apparatus	Racal
	Respirator Canister Model 456-00-07R 06	Racal
	Disposable respirators	Racal
	Respirator canister Model 110100	Survivair
	M-40A1 series masks	Tradeways (Md)
	Method for filtering CB agents from airflow in confined space	TSWG (R&D only)
	First responders mask (FIRM)	TSWG (R&D only)
Clothing		
	Mark IV permeable NBC Suit	ADI (UK)
	Remploy Tyvek F-M(ilitary) ensemble	ADI
	JLIST (Joint Service Lightweight integrated NBC protective suit technology)	CBDCOM (R&D only)
	STEPO (Self-contained toxic environment protective outfit)	Chemfab Corp (NH)
	Biomimetic materials	DARPA/Molecular Geodesics (R&D only)
	Man-in-Simulant Test Program	Dugway Proving Grounds (R&D only)
	Low-cost protective suits	Geomet Technologies
	Field Marking Kits	Giat Industries
	TOM suit kit	Giat Industries
	Gastight suit for internal breathing apparatus	Giat Industries

Type	Product	Location/PI
Clothing		
	UNISCAPH gastight suit for external BA	Giat Industries
	Cool Vest Personal Cooling Garment	ILC Dover, Inc.
	Chemturion: Reusable Level A Suit	ILC Dover, Inc.
	Ready 1 Limited Use Level A Suit	ILC Dover, Inc.
	Cooling Vests	Kappler Protective Apparel and Fabrics
	Responder CSM Garments	Kappler Protective Apparel and Fabrics
	Pressure test kits	Kappler Protective Apparel and Fabrics
	Chemical Protective Overgarment	Marine Corps Systems Command (R&D only)
	Functionally Tailored Fibers and Fabrics	Natick RDEC (R&D only)
	Firefighters Integrated protective Suit - Combat (FISC)	Natick RDEC (R&D only)
	Advanced Lightweight Chemical Protection	Natick RDEC (R&D only)
	Level B Suit	Responder-Geomet
	Level A Suit	Responder-Geomet
	SARATOGA-Pyjama Chemical Protective Undergarment	Tex-Shield, Inc (NJ)
	CW-66 Chemical Protective Flight Coverall	U.S. Air Force
	(BDO) Battledress overgarment	Winfield International (NY)
Clothing and Breathing		
	Domestic Preparedness Civilian PPE Testing Program	CBDCOM (R&D only)

Type	Product	Location/PI
Clothing and Breathing		
	(CBPSS) Chemical Biological Protective Shelter System	Engineered Air Systems (Mo)
	Individual Protective Kit	Giat Industries
	Rescue and Lifting Kit	Giat Industries
	Ventilated casualty bag	Giat Industries
	Ventilated casualty hood	Giat Industries
	ILC Dover Transportable Collective Protection System	ILC Dover
	M20A1 SCPE (Simplified Collective Protection System)	ILC Dover, Inc.
	Improvements to existing C/B Bomb suit	Tech Escort Unit (R&D only)
	Expedient Hazard Reduction System	TSWG (R&D only)
	ILC Dover Transportable Collective Protection System	ILC Dover
	Protection assessment test system	U.S. Army (R&D only)

Chemical Agent Detection

30-Sep-98

Availability	Product	Source or Laboratory/PI
Commercial		
	ACAMS Automated Continuous Air Monitoring System	ABB Process Analytics
	Remote molecular air quality monitoring system (FTIR)	AIL Diversified Products Group
	Nerve agent vapor detector (NAVD)	Anachemia Canada Inc.
	Paper, Chemical Agent Liquid Detectors, 3-WAY	Anachemia Canada Inc.
	C2 chemical agent detector kit	Anachemia Canada Inc.
	CUB 800	Bear Instruments, Inc.
	Infrared Detector for Gas Chromatograph	Biorad, Digilab Division
	Transform spectrometer	Bomem Inc.
	TYPE 1306 Toxic-Gas Monitor	Bruel and Kjaer Instruments
	TYPE 1302 multigas monitor	Bruel and Kjaer Instruments
	Rapid Alarm and Identification Device (AID-1)	Bruel and Kjaer Instruments
	Chemical Surety Chamber and Lab	Calspan Corporation
	Automatic Liquid Agent Detector (ALAD) System	Calspan Corporation
	Miniature Chemical Agent Monitor (MINICAM)	CMS Research Corporation
	Detector tubes	Draeger
	Continuous Chemical Agents Sensor (CHASE)	Elbit-Ati Instruments
	4100 vapor detector	Electronic Sensor Technology

Availability	Product	Source or Laboratory/PI
Commercial		
	Improved Chemical Agent Monitor (ICAM-APD)	Environmental Technologies
	APD2000 Hand-held CW/radiation detector/monitor	Environmental Technologies
	Miniature Chemical Agent Detector (ICAD)	Environmental Technologies
	Chemical Agent Monitor (CAM)	Environmental Technologies
	Detalac Automatic Detector of Nerve gas agents	Giat Industries
	Environmental Vapor Monitor (EVM)	Graseby Dynamics Ltd (UK)
	Point Chemical Agent Detector (GID-3)	Graseby Dynamics
	HP 6890 Plus Gas Chromatograph	Hewlett-Packard
	HP 2350 Atomic Emission Detector	Hewlett-Packard Co.
	HP 5973 MSD	Hewlett-Packard Co.
	Improved Chemical Agent Monitor (ICAM)	Intellitec (Florida)
	M8A1 automatic chemical agent alarm (M43A1 and M42)	Intellitec (Florida)
	AN/KAS-1 Chemical Warfare Directional Detector (standoff)	Intellitec (Florida)
	M21 Remote sensing chemical agent alarm (RSCAAL)	Intellitec (Florida)
	Nerve Agent Immobilized-Enzyme Alarm and Detector (NAIAD)	Jasmin Simtec Limited
	SAW MiniCAD miniature chemical agent detector	Microsensor Systems, Inc.
	CW Sentry	Microsensor Systems, Inc.,
	RCAD II Monitor	Microsensor Systems, Inc.,

Availability	Product	Source or Laboratory/PI
Commercial		
	EKHO	Mine Safety Appliances Co.
	Field MINICAMS (FM-2000)	OI Analytical, Inc
	Phemtochem Ion Mobility Spectrometer, Model 110	PCP Inc.
	API 365	Pe Sciex
	Microchromatography	PerSeptive Biosystems, Inc.
	AP2C Family of Chemical Agent Detectors	Proengin S.A. (France)
	MINIRAE Plus	Rae Systems
	Direct-Reading Tubes	Sensidyne
	M90D1-A Chemical detector	Sensor Application Inc (VA)
	Scentograph Plus II with AID/RCD Detector	Sentex Systems Inc.
	Scentoscreen (Gas Chromatograph) with Argon Ionization Detector	Sentex Systems Inc.
	SCX-20 VOC Monitor	Spectrex Corporation
	Dual Flame Photometric Detector	SRI Instruments Inc.
	TestMate portable blood cholinesterase test system	TestMate, Inc
	Miran Sapphire	The Foxboro Company
	Chemical agent point detection system for ships (CPDS)	Tradeways Ltd (MD)
	M8 Chemical detection paper	Tradeways Ltd (VA)
	M9 Chemical detection paper	Tradeways Ltd (VA)
	M256A1 Chemical detection kit	Truetech Inc (NY)
	M272 Water testing kit	Truetech Inc (NY)
	M18A2 Chemical detection kit	Truetech, Inc (NY)

Availability	Product	Source or Laboratory/PI
Commercial		
	SATURN	Varian Chromatography Systems
	Portable GC/MS systems: SpectraTrak and CADIS	Viking Instruments Corporation
	Mass-Spec-On-Chip (MSOC)	Westinghouse Electronic System
Field testing		
	FBI Fly-away Laboratory	Unknown
	Nonintrusive interrogation of closed containers	Battelle Columbus
	CB mass spectrometer (CBMS I)	Bruker Instruments
	Air Transportable Modular Analytical Laboratory (MODLAB)	CBDCOM
	XM22 Advanced Chemical Agent Detector/alarm (ACADA)	ERDEC/Nowack
	SOF Chemical Agent Detector w low false positives	Graseby
	GI-MINI Miniature Chemical Warfare Detector/Monitor	Graseby Dynamics
	Rugged, portable GC-MS for CW agents	LLNL
	Multipurpose integrated chemical agent alarm (MICAD)	Lockheed Martin
	Shipboard Automatic Liquid Agent Detector (SALAD)	Naval Sea Systems Command
	Improved Point Detection System (IPDS)	Naval Sea Systems Command
	SAWRHINO (nerve and vesicant agents)	NRL/Veneskey
	LabChip applications to ChE and immunoassays of selected CBW agents	ORNL/Ramsey

Availability	Product	Source or Laboratory/PI
Field testing		
	Bruker Ims Point Chemical Detector (PCD)	Bruker Saxonia Analytik (Germany)
	Minitube Air Sampling System (MASS)	Canadian Centre for Advanced Instrumentation
	Chemical Agent Detection System II (CADS II)	Scientific Instrumentation Limited (Canada)
	Laser Remote Detector	Research Institute 070 BRNO (Czech Republic)
	MARK 1 Detector Kit Chemical Agent Residual Vapor (RVD)	Richmond Packaging (UK) Limited
U.S. Government		
	Contract Labs	EPA Envir Response Team Center (NJ) (Lafornara)
	TAGA 6000E MS/MS-triple quadrapole	EPA Envir Response Team Center (NJ) (Lafornara)
	Mobile lab	EPA Envir Response Team Center (NJ) (Lafornara)
Laboratory Research		
	Canine detection of low level CW	Auburn: Lackland AFB
	Wearable dosimeter indicating cumulative exposure	CWC Treaty Lab
	Miniature GC-IMS	DARPA (Technispan) G. Spangler
	Joint service lightweight standoff chemical agent detector (JSLSCAD)	JPOBD
	Joint Service Chemical Agent Detector (JCAD)	JPOCD
	Joint Service Chemical Warning and Identification LIDAR	JPOCD
	Miniaturized liquid chromatography	M.D. Porter, Iowa State University

Availability	Product	Source or Laboratory/PI
Laboratory Research		
	Micro-mass spectrometer for containment gas monitoring	M.P. Sinha, Imaging and Spectrometry Systems Technology
	Neuron Chip	NRL/F. Ligler
	Micro unmanned chemical and biological sensor vehicle	NRL/R. Foch
	CB mass spectrometer Block 2 (CBMS II)	Oak Ridge National Laboratory
	Advanced Ion-trap mass spectrometry	ORNL/S. McLuckey
	Capillary electrophoretic methods for monitoring spacecraft water	NASA/R.L. Sauer
	Enzyme-FET	Sandia National Lab/Thomas
	Parallel Micro Separations-based Detection (PMSD)	SNL/Vitko
	Noninvasive methemoglobin measurement	USAMRMC
Military		
	GS/MS detection of chlorovinylarsenous acid (from Lewisite) in urine	USAMRMC
	GS/MS detection of phosphonic acids (from GA,VX) in patient fluids	USAMRMC

Epidemiology Resources

13-Oct-98

Availability	Product	Source
Operational		
	Associate	Carter-Melloy Corp
	National Electronic Telecommunication-System for Surveillance (NETSS)	CDC
	Public Health Lab Information System	CDC
	Communicable Disease Surveillance Center (CDSC)	England/Wales
	Salm-Net	European Union
	ProMed (Program for Monitoring Emerging Diseases)	FASEB
	National Antimicrobial Resistance Monitoring System (NARMS)	FDA/CDC/USDA
	GIDEON (Global Infectious Disease Epidemiology Network)	Gideon USA
	Canadian Bacterial Disease Network (CBDN)	NCE
	Air Force Global Surveillance	U.S. Air Force
	WHO Weekly Epidemiological Record (WER)	WHO
	Outbreak	WWW
	Emerging Infectious Disease Initiative	CDC
Planned		
	Global Emerging Infections Surveillance and Response System (DoD-GEIS)	DoD
	Global Public Health Intelligence	Health Canada

Symptom-Based Diagnosis Systems

30-Sep-98

Availability	Product	Source
Operational		
	Associate	Carter-Melloy Corp
	NBC indicator symptom matrix	Defense Protective Services/ M. Dougherty
	Global Infectious Disease and Epidemiology Network	Gideon, USA
Planned		
	Emergindex	Micromedex, Inc.
	Drugdex	Micromedex, Inc.
	Poisindex	Micromedex, Inc.

Biological Agent Detection

13-Oct-98

Availability	Product	Source or Laboratory/PI
Commercial		
	LIfeChip High-Density Nucleic Acid Microarrays	Affymetrix, Inc.
	Profile 1 Bioluminometer	Environmental Technologies
	SMART Biological Warfare Detection Tickets	Environmental Technologies
	Biological integrated detection system (BIDS)	Environmental Technologies
	LightCycler (LC32) Thermal cycler microvolume fluorimeter	Idaho Technology
	SpinCon High-volume Portable Air Sampler	Midwest Research Institute (MRI)
Field tests		
	PathoSeq bacterial gene sequence library	Affymetrix, Inc.
	Modification of NMRI handheld BW tickets	Battelle/NMRI
	CB mass spectrometer (CBMS I)	Bruker Instruments
	Biological Microchips for Field Analysis of Microorganisms	DARPA (ANL, A. Mirzabekov)
	Mini Mass Spectrometer for Biodetection	DARPA (JHU/APL, W. Bryden)
	16S Ribosomal RNA Hierarchical Analysis	DARPA (Northwestern U.) Stabl
	Miniature Environmental Air Sampler Using Aerogel	DARPA (PSR, Inc.) UVA, C. Daitch, P. Norris
	Hierarchical Analysis of Unknown Biological Samples	Duke Univ./Wilson

Availability	Product	Source or Laboratory/PI
Field tests		
	Simultaneous immunoPCR and genomic PCR	DuPont/Ebersole
	ALERT Lateral Flow Immunoassay Tickets	ERDEC/Emanuel and Valdes
	Integrated Virus Detection System	ERDEC/Wick and EnVirion, L.C.
	High speed chemical analysis of DNA by TOF-MS	GeneTrace Systems
	Interim biological agent detector (IBAD)	JPOBD
	Compact DNA-based Bacterial Identification by Flow Cytometry	LANL/B. Marrone
	Antibiotic Resistance Detection	LANL/Jackson; N. Ariz. U./ Keil
	BW Genetic Sequencing	LANL/Jackson; LLNL; Duke/Wilson
	Self-assembling thin film biosensors	LBNL/Charych
	Improved methods to isolate and process DNA from environmental samples	LLNL/Carrano; LANL/Kuske
	Miniature PCR-based Bioagent Detector	LLNL/Marriella
	Marine Corps Unit Biological Detector	MARCORSYSCOM/Bryce
	DNA Dipstick	Molecular Tools, Inc./ Boyce-Jacino
	Handheld Assay SMART Tickets	NMRI/Churilla
	Automated Programmable	Nanogen, Inc. Electronic Matrix APEX microchip
	Rapid PCR assays for BW detection	NMRI/Long and identification

Availability	Product	Source or Laboratory/PI
Field tests		
	Single Particle Fluoresence Counter	NRL/Eversole
	Anaylate 2000 Fiberoptic waveguide biodetector	NRL/Ligler
	LabChip applications to ChE and immunoassays of selected CBW agents	ORNL/Ramsey
	Long range biological standoff detection system (XM94)	Schwartz Electro-Optics
	Chem/Bio Sentry System	Tech Escort Unit
Laboratory		
	FBI Fly-away Laboratory	Unknown
	Liquid phase piezoelectric immunosensors	A.A. Suleiman, Southern U.
	Joint biological point detection system (JBPDS)	CBDCOM
	Nanoscale DNA	CuraGen Corp.
	Advanced Diagnostics Program	DARPA
	Unmanned Aerial Vehicle-Borne Hybrid Optical Sensor	DARPA (Electro-Optics, Inc.) P. Titterton
	Novel Biodetection by Spore-specific Phosphorescen	DARPA (IIT Research Inst.) K. Rajan
	Next Generation, Integrated Biosensor Research	DARPA (Pacific Sierra) UVA, C. Daitch, P. Norris
	Smart Aerogels for Application in Biowarfare	DARPA (PSR, Inc.) UVA, C. Daitch, P. Norris
	Upconverting Phosphor Flow Cytometer	DARPA (SRI, J. Carrico)
	Upconverting Phosphor Compact	DARPA (SRI, J. Carrico) Handheld Biosensor
	Novel Antibody Reagents (Immunoplastics) for Sensors	DARPA (U. of TX, G. Georgiou)

Availability	Product	Source or Laboratory/PI
Laboratory		
	Structure-based Ligands to Capture Microorganisms	DARPA (U. of Ala., Birm) L. DeLucas
	Capture of Pathogenic Microbes	DARPA (Utah State U.) L. Powers
	Pathogenic Microbe Sensor Technology	DARPA (Utah State U.) B. Weimer
	Rapid methods of detecting BW agents on food	FDA; U. of Md.
	Rapid bacterial testing for spacecraft water	G.A. McFeters, Montana State U.
	Microbial monitoring based on quantitative PCR	G.H. Cassell, U. of Alabama, Birmingham
	MALDI-MS for identifying intact whole bacteria	Joint Inst Food Safety and Appl Nutrition/Musser
	Joint Biological Remote Early Warning System (JBREWS)	JPOBD
	Simultaneous monitoring of multiple bacteria in spacecraft	M.D. Eggers, Genometrix, Inc.
	IGEN PCR Biosensor Assay	NMRI/Churilla
	Recombinant antibodies for BW Agents	NMRI/Churilla
	DNA Detection via Current-Rectifying Oligonucleotides	Northwestern U./Mirkin
	Force Amplified Biological Sensor (FABS)	NRL/Colton
	A Multiplexed Immunosensor based on Lateral Force Microscopy	NRL/Gaber
	Automated Multiagent Sensor	NRL/Ligler
	Neuron Chip	NRL/Ligler
	Micro unmanned chemical and biological sensor vehicles	NRL/R.Foch

Availability	Product	Source or Laboratory/PI
Laboratory		
	CB mass spectrometer Block 2 (CBMS II)	Oak Ridge National Laboratory
	Recombinant antibodies specific to *Bacillus anthracis* spores	ONR/LLNL (Leighton)
	Advanced Ion-trap mass spectrometry	ORNL/McLuckey
	Bioaerosol Detector System based on Aerogel	Pacific-Sierra Research Corp
	UV Fluoresence Detection of BW Agents on Surfaces	Sandia NL/Thomas
	Parallel Micro Separations-based Detection (PMSD)	SNL/Vitko, Thomas
	Taqman PCR-based BW assays	USAMRIID
	Automated Nucleic Acid Extractor	USAMRIID/Xohox, Inc
	Deployable diagnostic kit for biowarfare agents	USAMRMC

Decontamination Products

30-Sep-98

Availability	Product	Source/Location
Commercial		
	M11, M13 Man-portable decontamination application systems	All-Bann Enterprises/ Tradeways Ltd (MD)
	M12 Powered vehicle-mounted multipurpose decontaminating apparatus	All-Bann Enterprises/ Tradeways Ltd (MD)
	DS2P Decon solution	All-Bann Enterprises/ Tradeways Ltd (MD)
	M17 Lightweight decontamination system (Sanator)	Engineered Air Systems (MO)
	Emergency Response Equipment Package	HAZ/MAT DQE Inc (IN)
	Hospital-based Decontamination Equipment Package	HAZ/MAT DQE Inc (IN)
	Transportable Decontamination Systems	Modec Inc. (Denver)
	Decontamination Kit No. 2	Tradeways Ltd
	M291 Decontamination kit for individual equipment	Tradeways Ltd
	M258 Skin decontamination kit	Tradeways Ltd
	STB super tropical bleach	Unknown
Field tests		
	XM21/XM22 Modular decontamination system	CBDCOM
	Wound decon systems	USAMRIID
	Mediclean Spray/Suction Units	Karcher (Germany)
Research		
	Non-toxic, non-corrosive enzyme-based foam decon system	Arthur D. Little/Altus Biologics

Availability	Product	Source/Location
Research		
	Sorbent decontamination system	CBDCOM
	Sensitive equipment decontamination	CBDCOM
	Biomimetic materials	DARPA/Molecular Geodesics, Inc. (D. Ingber)
	Molecular decoys	DARPA/U. of Michigan (Baker)
	Enzymatic Decontamination	ERDEC/DeFrank
	Solid state absorber/oxidizers for CW decon	LANL/Earl
	Ozone based methods for CW decon of equipment	LANL/Earl
	Fenton chemistry (peroxides) for CW and BW decon	LANL/Earl
	Low temperature plasma jet	LANL/Earl
	Oxidizing solutions for CW decon of equipment and property	LLNL
	Gel carrier for vertical surfaces	LLNL/Raber
	Surfactant-based Decontamination Solution	ONR/NSWC Dahlgren (Brown)
	Quaternary Ammonium Complex Decontaminant	ONR/NSWC Dahlgren (Cronce)
	Chemical/UV decontamination method	Optimetrics, Inc
	Hydrolyzing foams	Sandia National Lab/Zelikoff
	Corona discharge air purification technology	SRI International
	Fixed site decontamination system	U.S. Marines
	Lightweight portable decontamination system	U.S. Marines

Chemical Agent Treatments

07-Oct-98

Agent	Treatment	Source	Availability
Cyanide			
	Amyl nitrite+sodium nitrite+ sodium thiosulphate	Pasadena	Commercial
	alpha adrenergic antagonists	Other	Commercial
	Superactivated charcoal	Other	Commercial
	4-dimethylaminophenol (4-DMAP)	Germany	Foreign
	Kelocyanor (dicobalt EDTA)	Germany	Foreign
	Hydroxocobalamin (vitamin B12a)	France	Foreign
	Stroma-free hemoglobin	USAMRICD	IND
	p-aminooctanoylphenone (PAOP)	USAMRICD	Preclinical
	p-aminoheptanophenone (PAHP)	USAMRICD	Preclinical
	p-aminopropiophenone (PAPP)	USAMRICD	Preclinical
	8-aminoquinoline derivatives	Other (Steinhaus et al.)	Preclinical
	Alpha-ketoglutaric acid	Indian Defense R&D Establishment	Preclinical
Nerve agents			
	Pralidoxime chloride (2-PAM)	Meridian Med Tech; Quad Pharm; Wyeth	Commercial
	Diazepam	Abbott, Lederle, Parke-Davis, others	Commercial
	Pyridostigmine bromide	ICN Pharmaceuticals	Commercial
	Atropine	Meridian Med Tech: Kalli DuPhar; 3M-Reiker	Commercial

Agent	Treatment	Source	Availability
Nerve agents			
	Obidoxime	Czech Republic military	Foreign
	Reactive topical skin protectants	USAMRICD	Preclinical
	Methanesulphate salt of pralidoxime	UK	Preclinical
	Nicotine hydroxamic acid methiodine	USAMRICD	Preclinical
	Monoisonitroacetone (MINA)	USAMRICD	Preclinical
	Butyrylcholinesterases (horse, human, mutants)	USAMRICD	Preclinical
	Acetylcholinesterase	USAMRICD	Preclinical
	Catalytic monoclonal antibodies	USAMRICD	Preclinical
	Memantine	USAMRICD	Preclinical
	NMDA receptor blockers	USAMRICD	Preclinical
	Pro-2-PAM	Unknown	Preclinical
	H series of oximes	Czech Republic military and others	Preclinical
	Carboxylesterase	USAMRICD	Preclinical
Phosgene			
	Oxygen+ventilation+ bronchodilators	Other	Commercial
	Aminophylline	Other	Commercial
	Hexamethylene tetramine (HMT)	Other	Commercial
	Corticosteroids	Other	Preclinical
	Cysteine (and N-acetylcysteine)	USAMRICD	Preclinical
Vesicants			
	Soap and water	Other	Commercial

Agent	Treatment	Source	Availability
Vesicants			
	Hypochlorite solution (<1%)	Chlorox	Commercial
	British Anti-Lewisite (dimercaprol)	Becton Dickinson Microbiology Systems	Commercial
	M258A1 Decon kit	Tradeways, Ltd. (MD)	Commercial
	Nitric oxide synthase inhibitors	Canadian Defence Research Establishment	Preclinical
	Dexamethasone/heparin/ promethazine combos	Other	Preclinical

Biological Agent Treatments

07-Oct-98

Agent	Treatment	Source	Availability
Anthrax			
	Ciprofloxacin	Bayer	Commercial
	Anthrax Vaccine	Michigan Biological Products	Commercial
	Doxycycline	Parke-Davis; Pfizer; Lederle	Commercial
	Biostructure Mapping by STEM, Cryo-EM, EELS, and SPM: Anthrax Toxin	Brookhaven NL/Furlong	Preclinical
	Structure-based Drug Design for Microorganism-associated Proteins	DARPA/UAB (DeLucas)	Preclinical
	Anthrax Toxin Structure and Function	NIDR/Leppla	Preclinical
	Control of Protective Antigen Synthesis by B. anthracis.	UTex Health Ctr Houston/Koehler	Preclinical
Bacteria			
	Novel Broad Spectrum Antimicrobial Agents	DARPA/Isis Pharmaceuticals (D. Ecker)	Preclinical
	Novel Broad Spectrum Antimicrobial Agents-Gene Expression	DARPA/ SmithKline Beecham (M. Rosenberg)	Preclinical
	Novel Targets of Pathogen Vulnerability	DARPA/ Stanford U. (L. Shapiro)	Preclinical
	Sequential Auto Vaccination by Stem Cells	OSIRIS Therapeutics (D. Marshak)	Preclinical

Agent	Treatment	Source	Availability
Brucella			
	Rifampin (Rifadin)	Merrell Dow Pharmaceuticals	Commercial
	Recombinant Brucella Vaccine Development	(USSR)/Noskov; USAMRIID/ Friedlander	Preclinical
	Immunogenicity of Recombinant Brucella Abortus Proteins	LSU Med Ctr/ Roop	Preclinical
C. Botulinum			
	Immune globulin from human donors	(Frankovich and Arnon)	IND
	Vaccine (toxoids A-E)	CDC	IND
	Trivalent botulinum antitoxin (A,B,E)	CDC	IND
	Horse antibotulism serum (globulin)	USAMRIID	IND
	Botulinum Vaccine	USAMRIID	IND
	Mechanism of Botulinum Toxin Action	Thomas Jefferson Univ./Simpson	Preclinical
	Aminopyridines (3,4-diaminopyridine)1	USAMRIID	Preclinical
	Monoclonal antibodies	USAMRIID	Preclinical
	Recombinant vaccines	USAMRIID	Preclinical
	Chimer of botulinum toxin receptor-binding protein	USAMRIID	Preclinical
Dengue			
	Functional Analysis of Dengue Virus Antigens NS3 and NS5	UKansas Med Center/ Padmanabhan	Preclinical
	Mechanisms of Immunopathology in Dengue Hemorrhagic Fever	UMass Med Ctr/ Ennis	Preclinical
Ebola			
	Immunologic and Epidemiologic Studies of Emerging Viruses	Scripps Res Inst/ Buchmeier	Preclinical

Agent	Treatment	Source	Availability
EEE			
	EEE Vaccine	Unknown	IND
	Acute Alphavirus Encephalitis	Johns Hopkins SPH/Griffin	Preclinical
Lassa			
	Immunologic and Epidemiologic Studies of Emerging Viruses	Scripps Res Inst/ Buchmeier	Preclinical
	Molecular Basis of Arenavirus Virulence	U Wisconsin/ Salvato	Preclinical
Multiple			
	Naked DNA/gene gun vaccines	USAMRIID	Preclinical
	Multiagent replicon vaccines	USAMRIID	Preclinical
Plague			
	Plague Vaccine	Greer	Commercial
	Streptomycin	Lilly; Pfizer	Commercial
	Doxycycline	Parke-Davis; Pfizer; Lederle	Commercial
	Plasmid PCD-Encoded Virulence Determinants in Plague	U Kentucky/ Straley	Preclinical
	Mechanism of Bacterial Metastasis in Plague	UMass Med Sch/Goguen	Preclinical
Q-Fever			
	Tetracyclines	Numerous drug companies	Commercial
	Q-Vax	Australian product	Foreign
	Q-Fever Vaccine	IND 610	IND
	Pathogenic Roles of Coxiella burnetti Surface Proteins	Texas A&M/ Samuel	Preclinical
	Surface Change and Virulence in Coxiella burnetti	Wash. State Univ./Mallavia	Preclinical

Agent	Treatment	Source	Availability
Ricin			
	alpha deglycosylated A chain as antigen	USAMRIID	IND
	Antiricin rabbit antibodies	USAMRIID	Preclinical
	Formalin treated toxoid	USAMRIID	Preclinical
	Toxoid in galactide-glycolyde	USAMRIID	Preclinical
SEB			
	Staphylococcal Toxins	Kansas State/ Iandolo	Preclinical
	Immunosuppressive Action of Staphylococcal Enterotoxins	Temple Univ Sch of Med/Rogers	Preclinical
Smallpox			
	Cidofovir (Vistide)	Giliad Pharm	Commercial
	Smallpox Vaccine (Dryvax)	Wyeth	Commercial
	Smallpox Vaccine (DoD)	USAMRIID	IND
	Vaccinia DNA Replication: Genetics and Molecular Biology	Cornell Univ Med Coll/Traktman	Preclinical
	Viral Inhibition of Host Defenses	Duke Univ Med School/Pickup	Preclinical
T-2 Mycotoxin			
	M238A1 skin decon kit	USAMRICD	
	Multi Shield TSP barrier cream	Interpro, Inc (Mass)	Commercial
	Superactive activated charcoal	Other	Commercial
	Corticosteroids (systemic)	Other	Commercial
	XE-555 resin (M291 decon kit)	Tradeways Ltd (MD)	Commercial
	Mycotoxin with carrier protein	USAMRIID	Preclinical
	BN52021	USAMRIID	Preclinical
	Prophylactic enzyme induction	USAMRIID	Preclinical

Agent	Treatment	Source	Availability
T-2 Mycotoxin			
	Despeciated monoclonal anti-idiotype antibody	USAMRIID	Preclinical
Toxins			
	Red Blood Cell Pathogen Defense-Destruction	DARPA/Boston U (M. Bitensky)	Preclinical
	Polyvalent Inhibitors of Microorganisms, Viruses, and Toxins	DARPA/Harvard U (G. Whitesides)	Preclinical
	Structural Biology of Bacterial Toxins	DARPA/Los Alamos National Lab (G. Gupta)	Preclinical
	Intracellular Sensors of Virulence	DARPA/U. of Michigan (R. Kopelman, et al.)	Preclinical
	Instant Immunization	DARPA/U.TX-South Western Med. Ctr. (S. Johnston)	Preclinical
	Red Blood Cell Pathogen Defense-Decoy	DARPA/UVA (R. Taylor)	Preclinical
Tularemia			
	Streptomycin	Lilly; Pfizer	Commercial
	Tularemia Vaccine	Unknown	IND
VEE			
	C-84 VEE Vaccine	USAMRIID	IND
	TC-83 VEE Vaccine	USAMRIID	IND
	Structure-based Drug Design for Microorganism-associated Proteins	DARPA/UAB (DeLucas)	Preclinical
	Molecular Evolution of Guanarito Virus	Southwest Fnd for Med Res/ Rico-Hess	Preclinical
	In-vitro Construction of Attenuated VEE Virus Mutants	UNC Chapel Hill/Johnston	Preclinical

Agent	Treatment	Source	Availability
Virus			
	Ribavirin (Virazole)	ICN Pharmaceuticals	Commercial
	Developmental Proteins to Prevent Human Injury from Pathogens	DARPA/enVision (E. Barnea)	Preclinical
	Super Immune Cells	DARPA/Harvard Med School (D. Scadden)	Preclinical
	Novel Bacteriophage Therapies for Vibrio Cholerae Infection	DARPA/Harvard U. (J. Meklanos)	Preclinical
	Invasive (Intra-cellular) Antibodies	DARPA/Scripps Research Inst. (P. Ghazal)	Preclinical
	Heat Shock Protein-Peptide Complexes as Anti-Viral Agents	DARPA/U. of Connecticut (P. Srivastava)	Preclinical
	Structure-based Design of Acute Countermeasures to Viruses	DARPA/U. of TX at Galveston (R. Shope)	Preclinical
	Prevention of Virus Assembly in Host Cells	DARPA/U. of Wisconsin (S. Kornguth)	Preclinical
	Cytotoxic T Cell Responses to Virus Infection	Scripps Research Inst./Whitton	Preclinical
	Glycyrrhizic acid derivatives	USSR/Pokrovsky; USAMRIID/ Huggins	Preclinical
	Monkeypox Virus Genome Sequencing	USSR/ Shchelkunov; USAMRIID/ Jarling	Preclinical
WEE			
	WEE Vaccine	Unknown	IND

Prevention and Treatment of Psychological Effects

07-Oct-98

Source/Location	Product	Focus	Availability
American Psychiatric Assoc	Disaster Psychiatry Web Site	Multiple	Open Literature (www)
	Committee on Psychological Responses to Disaster	Multiple	Open Literature (www)
American Psychological Assoc	Disaster Response Network	Multiple	Open Literature (www)
American Red Cross	Disaster Mental Health Services	Multiple	(Training, short-term intervention)
Department of Veterans Affairs	On-line Publications, database	Victims	Open Literature (www)
	National Center for PTSD	Victims	Open Literature (www)
Disaster Mental Health Institute	Training, consultation, interventions	Multiple	Fee-for-Service
International Critical Incident Stress Foundation (Mitchell)	Critical Incident Stress Debriefing training, network of providers	Workers	Fee-for-Service
International Society for Traumatic Stress Studies	Web Site, journal on stress and coping	Victims	Open Literature (www)
National Research Council	Studies on Risk Communication, 1989, 1996	Community	Open Literature (www)
Rutgers Center for Environmental Communication (Chess)	Studies, advice to governments and industry on dealing with public concern	Community	Fee-for-Service
Substance Abuse and Mental Health Services Admin (Flynn)	FEMA Crisis Counseling Assistance and Training Program (CCP)	Victims	Federal Response Plan

Source/Location	Product	Focus	Availability
U of Delaware Disaster Research Center (Nigg)	Disaster Recovery as a Social Process and similar studies	Multiple	Open Literature (www)
Uniformed Services Univ of Health Sciences (Norwood)	Center for Stress Studies— Studies, advice on stress and coping in military	Multiple	Open Literature (www)
Uniformed Services Univ of Health Sciences (Ursano)	Psychiatry Dept.—Studies, advice on stress and coping in military situations	Multiple	Open Literature (www)
Walter Reed Army Institute of Research (Belenky)	Studies, advice on stress and coping in military situations	Multiple	Open Literature (www)

Computer Models

30-Sep-98

Availability	Product/Model	Agent Type	Source
Beta testing			
	BWD Incident Repository	Bio	DARPA/Oracle (S. Kennedy)
	BWDAD (Biological Warfare Defense Anchor Desk)	Bio	DARPA/SAIC (R. Goodwin)
	BITLAS (Biological integration team large area simulation model)	Bio	OptiMetrics, Inc.
	Accelerated Consequences Management	C/B	DARPA (J. Silva)
	GRIP (Global Response Incident Planner)	C/B	DARPA/BBN (M. Callaghan)
	Field Inventory Survey Tool	C/B	DARPA/BBN (M. Callaghan)
	Casualty Triage Tag	C/B	DARPA/Ellora Software (J. Bachant)
	MMTandE (Military Medical Training and Evaluation)	C/B	DARPA/Michigan S.U. (J. Downs); U of TX (S. Hufnagel); SAIC
	EMCR (Electronic Medical Care Record Repository)	C/B	DARPA/Oracle (S. Kennedy)
	Essential Medical Data Set	C/B	DARPA/Oracle (S. Kennedy)
	CODA (Chemical/biological Operational Decision Aid)	C/B	DARPA/Pacific - Sierra Research
	COC (Command Operations Center of the Future)	C/B	DARPA/ScrenPro (J. Mantock)
	AAHAWS (Automated atmospheric hazard assessment/warning system)	C/B	Mevatec Corp/ ENSR Consulting

Availability	Product/Model	Agent Type	Source
Operational			
	HASCAL/SCIPUFF (Hazard Assessment System for Consequence Analysis)	C/B	Defense Special Weapons Agency
	CATS/WMD (Consequences Assessment Tool Set)	C/B	Defense Special Weapons Agency
	NBC Warn (Nuclear, biological, and chemical warning and reporting network software)	C/B	OptiMetrics, Inc.
	ALOHA	Chem	EPA
Planned			
	Pgm for Response Options and Technology Enhancements for Terrorism in Subways	C/B	Argonne National Lab (Policastro)
	Urban Transport of CW/BW Aerosols	C/B	Lawrence Livermore (Ermak, Imbro); McArthur Found. (Stanford/Wilkening)
	CBW Environment/Challege and Mobile Force Operability Modeling and Simulation	C/B	NSWC Dahlgren

GAP AND OVERLAP ANALYSIS

A wide variety of sources were used in assembling the above inventory. The initial meeting of the committee in July of 1997 provided an overview of important organizations and R&D programs within the federal government. Follow-up with the briefers provided a more detailed list of projects and points of contact for technical information. The Office of Emergency Preparedness shared information on promising technology from its files, and of course the committee members themselves contributed both personal contacts and specific information from their own files and experience. The World Wide Web provided much information about both relevant commercial products and R&D activity, and the following databases were accessed and searched: National Technical Information Service, Defense Technical Information Center, Federal Research in Progress, Federal Conference Papers, Medline, MedStar, HSRProj.

Although we are still actively seeking additional information on many of the technologies already located, information on the products in the above inventory was distilled from a ProCite database of more than 430 records and entered into a series of databases, a description of which constitutes this gap and overlap analysis. In the process, we eliminated most products or R&D that did not explicitly address military chemical or biological agents or appear to be sufficiently generic in nature to encompass those agents without a major change. Exceptions were made only in categories in which there were very few or no products or R&D explicitly directed at chemical and biological weapons. We also excluded technology represented in our database by only a single experiment, journal article, or SBIR contract (i.e., we focus on products and R&D *programs*).

The overall organization of the inventory roughly parallels that of the interim report: Separate sections address detection (in the environment, and in patient fluids), detection of a covert attack in a population (Epidemiology), protection, decontamination, treatment, psychological effects, and computer software. The inventory has no sections on pre-incident intelligence or safe and effective patient extraction, because we uncovered no relevant products or research (we recognize that there is a great deal of intelligence activity devoted to prevention of terrorism, but our task is to address consequence management—our inclusion of a pre-incident intelligence section in the interim report was solely to make the point that whatever the readiness of the civilian medical community, any pre-incident warning will amplify effectiveness manyfold). An additional difference from the interim report organization is a section on computer models. The inventory includes a source for the products or the laboratory and PI performing the research, and a judgment about the product's state of development (availability).

Detection

With 173 entries in the detector database, it became more manageable to divide the database into those detection devices intended for biological agents and those intended for chemical agents (there are 7 devices intended for both biological and chemical agents and these were included in each separate database).

Biodetectors

Most of the funding for biodetection devices comes from the Department of Defense (56%), with 18% from commercial ventures. DoE, FDA, NASA, and TSWG account for the remaining 26%. With only 6 (out of 73) devices commercially available, 92% are in either the field testing stage (40%) or still in the laboratory (52%).

Where they are used. There are only 17 devices in the database that are explicitly intended for diagnostic purposes, that is, detecting biological agent in fluid or tissue samples from a patient. Most (85%) current devices are designed to detect biohazards in the environment (liquid, air, surface, or other). Seven devices in the inventory are designed to detect agent in either patients or the environment, and numerous others aimed at environmental monitoring or detection could be adapted to patient diagnostics, but not without considerable additional research.

What is needed. The most prevalent medium needed is liquid (44%), although 18 devices are designed to detect agent in the air. Twelve devices utilize either liquid or air samples. Twenty-eight items (40%) provide numeric estimates of agent concentration. A third (33%) of the biodetection devices do not provide a quantitative estimate of the pathogen detected, and another 27% of the devices provide no information whatsoever about quantification.

Speed and portability. According to the inventory, device portability is evenly distributed among hand-held, carriable by man, truck-loaded, or fixed. However, much of the newest research focuses on miniaturization of detectors. Fifty-nine percent of the devices in the inventory will provide results in a matter of minutes. Eight devices (11%) can or will detect agent in a matter of seconds.

How they work. There are basically two types of technology needed in a biodetection device: (1) detection technology and (2) reporting technology. Detection technology refers to the mechanism by which the device differentiates the target from other organisms or molecules. Reporting technology refers to the transduction mechanism that makes the detection event apparent to a human observer. Thirty percent of the devices in the inventory depend upon nucleic acid hybridization for detection, while 23% use antibody/antigen binding. The remaining devices use chemical reactions, the composition of agent (size, charge, mass), ligand/receptor binding, or more than one of these technologies. Forty-one percent of the reporting technology is optical, with other devices using technologies based on charge, color, mass, electrochemical reaction, or some combination.

Chemical Detectors

There are 100 entries in the chemical detector inventory. Twenty-eight percent of the entries are funded by the Department of Defense and 56% by commercial companies. Other funders include DoE, EPA, NASA, and TSWG. Chemical detection devices are much more developed than their biological counterparts; 60% of the items in the inventory are commercially available, with only 13% still in the field testing stage and 16% in the laboratory. It is also worth noting that there are three commercial devices that are designed specifically for a civilian market.

Where they are used. The overwhelming majority (96%) of the chemical agent detectors are intended to detect agent in the environment, with only 4% designed to detect agent in patients.

What is needed. Forty-two percent of the devices provide a numerical estimate of agent concentration, but 47% only indicate the presence or absence of agent. There are four items in the inventory that will indicate a "High" or "Low" concentration of agent.

Speed and portability. Ninety-two percent of the chemical detection devices in the inventory are able to provide information about agents within minutes or seconds (43% and 49%, respectively). Fifty-one percent are hand-held devices, 10% can be moved by one man, 12% can be moved by truck to the site of a suspected attack, and 23% are fixed in one location (e.g., a ship or a laboratory).

How they work. The detector technology used by 24% of these devices depends upon a chemical reaction. Other technologies used in the detection process are: agent composition (mass, charge, or size) absorption, ligand/receptor binding, mass (mass spectrometry, piezoelectric, surface acoustic wave, or multiple technologies). For the reporting phase of the sources, technologies include: charge (1%), color (12%), electrochemical

(8%), atomic emission spectrum (1%), photo-acoustic (2%), surface acoustic wave (7%), or some combination or hybrid (8%) of these technologies.

Recognition of Signs and Symptoms in Patients (Diagnosis)

There are six products in this database. Three (the NBC indicator symptom matrix, Associates diagnostic software, and the Global Infectious Disease and Epidemiology Index [GIDEON]) are fully operational. The three Micromedex products (Poisindex, Drugdex, and Emergindex) are databases in use in poison centers and hospitals. Poisindex and Drugdex provide information on poisonous chemicals and drugs, while Emergindex is used for emergencies of unknown etiology. Only Emergindex is currently structured to provide diagnostic and treatment information based on signs and symptoms, but Micromedex is attempting to reengineer the other two databases to make this possible. At present, they require chemical or drug names as input.

The NBC indicator symptom matrix assumes that one of the traditional military chemical weapons is involved, and simply facilitates differential diagnosis among them. The other databases are larger in scope, but include some or all of the chemical or biological weapon agents.

Epidemiological Tools

The products in this database are potentially relevant in identifying outbreaks of disease in populations (as opposed to individual patients). There are 15 entries, 12 of which are operational at this time, the Emerging Infectious Disease Initiative of the CDC, which is a long-term project just getting under way, and the Global Public Health Intelligence of Health Canada, and the DoD's Global Emerging Infections Surveillance and Response System (DoD-GEIS) are two recently announced initiatives to be started in the near future.

Personal Protective Equipment

There are 63 entries in the personal protective equipment database. Of these, the vast majority (86%) of these products are designed to protect against both chemical and biological agents (8 are for chemicals only and one is for biological agents only). Many of the products are commercially available (44%), but 3 items are unique to the military. Also represented is equipment from 18 other countries. The U.S. Department of Defense (17 entries) and the multi-agency Technical Support Working Group (6 entries) sponsor laboratory research or field testing in this area. The type of equipment is evenly divided between protective clothing and breathing apparatus (both at 41%), with 11 entries that offer both types of protection.

Decontamination

Seventy-eight percent of the 33 products in this database involve strictly chemical decontamination. The remaining products are designed to decontaminate either biological agents alone (13%), or both biological and chemical agents together (9%). Only 10 (31%) are commercially available. Three of these 10 items are focused on decontamination of people, 4 on equipment or materiel decon, and 3 might be used for decon of either people or inanimate objects. Twenty-two entries (69%) are currently in research and development, which is largely funded by government agencies. The Department of Defense funds 50%, the Department of Energy 32%, and the multiagency Technical Services Working Group 9% of the decontamination products listed as in research or field testing. Only 4 of the 22 R&D items in the inventory are focused on biological agent decontamination; 2 items pertain to both chemical and biological agents; and 13 focus on chemical agents. Thirty-six percent of the products being researched are potentially applicable to human decontamination; the remainder focus solely on decon of inanimate objects.

Treatment

Of the 128 treatment products in the inventory, 88 (69%) are intended for biological agents, leaving 40 (32%) for the treatment of chemical agents. Funding for treatment research is provided largely by DoD (43%) and commercial institutions (34%). NIH accounts for 18% of the funding, leaving only 5% of the funding from the Public Health Service.

Biological Agents

The biological agents for which at least one treatment is being tested or is already available are: anthrax, brucella, *C. botulinum*, dengue, Ebola, EEE, Lassa, plague, Q-fever, ricin, SEB, smallpox, T-2 mycotoxin, tularemia, VEE, and WEE. Other entries involve broader treatments of more than one bacteria, virus, or toxin. Treatments for viruses, *C. botulinum*, and T-2 mycotoxin account for 35% of the treatment entries in the inventory (13%, 12%, and 10%, respectively)

Despite the abundant research on treatments of biological agents, most (65%) are in the preclinical stage of development. There are 13 (15%) INDs and 16 (19%) commercially available treatment products. There are 5 INDs for *C. botulinum* treatments, 1 for EEE, 1 for Q-fever, 1 for ricin, and 1 for smallpox. With the exception of tularemia, which has only one commercially available product and none in development, all of the agents listed above have at least one preclinical product under investigation. As might be expected, given the research status of most of the entries, there is no evidence or only partial evidence of efficacy in 53% of the treatment products. There is evidence of efficacy in animals in 21% of the entries, but only 3 (4%) entries with proven efficacy in humans.

Chemical Agents

The chemical agents considered for this inventory are: cyanide, nerve agents, phosgene, and vesicants. Out of the 40 treatment products in the inventory, 43% are for nerve agents, 30% for cyanide, 15% for vesicants, and 12% for phosgene. There is currently only one IND and it is for a cyanide treatment. As in biological treatments, most (53%) treatment products for chemical agents are in the preclinical stage of development; however, 35% of the chemical agent treatments are commercially available in the United States. There is animal evidence of efficacy in 68% of the entries, and 8 (20%) proven treatments in humans. The remaining products have no evidence or only partial evidence of efficacy.

Psychological Effects

This is a unique section of the inventory because the committee was unable to identify any "products" specifically connected with chemical or biological terrorism. The inventory thus focuses on information and resources regarding the psychological effects and treatment of trauma and disasters in general. There are 16 entries ranging from Web sites, to current studies, to publications. One entry focuses solely on rescue and health care workers; 4 solely on trauma victims themselves; and 2 focus on community-wide effects. The remaining seven include more than one of the above in their scope—usually victims and workers. Unfortunately, there is a dearth of information and resources about specific populations of victims such as the elderly, children, the disabled, and other special groups outside of the average adult male and female.

Computer Models

The 20 items in this database fall in two main categories: (1) information about agent transport, and (2) information about incident management. There are 10 models in each category. Most (13) are in beta testing, 4 are available for use at this time or are being used for purposes other than assisting authorities plan for responding to chemical or biological terrorism, and three are in the planning stage. The Department of Defense is funding 15 products, DoE 3, EPA 1, and 1 is funded by a commercial organization.

APPENDIX
C

Lethal and Incapacitating Chemical Weapons

Lethal and Incapacitating Chemical Weapons

Agent	Effects	Onset	First Aid
Nerve Agents			
GA (Tabun), ethyl N,N-dimethyl-phosphoramidocyanidate	Miosis, rhinorrhea, dyspnea, convulsions	seconds to minutes	Atropine, pralidoxime, anticonvulsants, ventilation
GB (Sarin), isopropyl-methylphosphorofluoridate			
GD (Soman), Trimethylpropylmethylphosphorofluoridate			
GF, cyclohexyl-methylphosphorofluoridate			
VX, o-ethyl S-[2-(diiospropylamino)ethyl]methylphosphorofluoridate			
Vesicants			
H, HD (mustard), bis-2-chlorethyl sulfide	Erythema, blisters, eye irritation, dyspnea	minutes to hours	Decontamination, topical antibiotics, bronchodilators, ventilation
CX (Phosgene oxime), dichloroformoxime			
L (Lewisite), J-chlorovinyldichloroarsine			British antilewisite
Cyanide			
AC (Hydrocyanic acid)	Loss of consciousness, convulsions, apnea	minutes	Nitrites, sodium thiosulfate
CK (Cyanogen chloride)			
Pulmonary Agents			
CG (phosgene), carbonyl chloride	Dyspnea, coughing	minutes to hours	Oxygen, ventilation, bronchodilators
DP (Diphosgene), trichloromethylchlorformate			

Continued on next page

Lethal and Incapacitating Chemical Weapons (*Continued*)

Agent	Effects	Onset	First Aid
Riot Control Agents CS, 2-chlorobenzylidenemalononitrile CN, 1-chloroacetophenone CR, dibenz (b,F)-1:4-oxazepine CA, bromobenzylcyanide DM, diphenylaminearsine DA, diphenylchlorarsine DC, diphenylcyanoarsine	Burning stinging of eyes, nose, airway, vomiting	seconds to minutes	Water for eyes, skin, bronchodilators, oxygen, ventilation for lungs
Opioids Carfentanil Sufentanil	Dyspnea, ataxia, catatonia	minutes	Naloxone, nalmefene, naltrexone, ventilation
Anesthetics Chloroform Halothane Cyclopropane Ether	Analgesia, loss of reflexes and conciousness	minutes	Ventilation
Antimuscarinics BX, 3-quinuclidinyl benzilate	mydriasis, erythema, ataxia, delirium	1 hour	Tacrine, physostigmine
Cholinergics Nicotine Epibatidine Anatoxin A	Weakness, tremors, apnea, convulsions	minutes	Mecamylamine

APPENDIX
D

Centers for Disease Control and Prevention List of Restricted Agents

VIRUSES

Crimean-Congo hemorrhagic fever virus
Eastern equine encephalitis virus
Ebola viruses
Equine morbillivirus
Lassa fever virus
Marburg virus
Rift Valley fever virus
South American hemorrhagic fever viruses (Junin, Machupo, Sabia, Flexal, Guanarito)
Tick-borne encephalitis complex viruses
Variola major virus (smallpox virus)
Venezuelan equine encephalitis virus
Viruses causing hantavirus pulmonary syndrome
Yellow fever virus

SOURCE: Centers for Disease Control and Prevention. *Additional Requirements for Facilities Transferring or Receiving Select Agents*, 42 CFR Part 72/RIN 0905-AE70. Atlanta: United States Department of Health and Human Services, 1997.

BACTERIA

Bacillus anthracis
Brucella abortus, Brucella melitensis, Brucella suis
Burkholderia (Pseudomonas) mallei
Burkholderia (Pseudomonas) pseudomallei
Clostridium botulinum
Francisella tularensis
Yersinia pestis
Coxiella burnetii
Rickettsia prowazekii
Rickettsia rickettsii

FUNGI

Coccidioides immitis

TOXINS

Abrin
Aflatoxins
Botulinum toxins
Clostridium perfringens epsilon toxin
Conotoxins
Diacetoxyscirpenol
Ricin
Saxitoxin
Shigatoxin
Staphylococcal enterotoxins
Tetrodotoxin
T-2 toxin

Index

A

Acceptable exposure limit, 37, 52, 56
Acetylcholinesterase (AChE), 60-61, 113, 115, 146, 148, 169
   See also Cholinesterase
Acoustic detectors, 53, 55, 57, 58, 59, 84, 256-257
Acquired immune deficiency syndrome, *see* Human immunodeficiency virus
Adenosine 5'-triphosphate, 87
Advance Lightweight Chemical Protection program, 38
AEL, *see* Acceptable exposure limit
Aeromedical Isolation Team, 26
Aerosols and vapors, 2, 22, 162
   atropine, 113
   biologic agents, 137, 138, 140, 146, 152, 154
   covert exposure, 41
   decontamination, 106
   detection and measurement, 44, 45, 47-51, 56, 86-87, 88-89, 226, 227
   personal protective equipment, 35, 38
   vesicants, 122, 123
   see also Respirators
Aged persons, 103, 108
Agency for Toxic Substances and Disease Registry, 75-76, 101

Agriculture
   chemicals industry, 39, 116
   crop contamination, 22
AIDS, *see* Human immunodeficiency virus
Aircraft, *see* Unmanned aerial vehicles
American Psychiatric Association, 171, 250
American Psychological Association, 171, 250
American Red Cross, 167, 171, 250
American Society for Testing and Materials, 35
Aminophenones, 127
Aminopyridines, 151, 242
Amyl nitrite, 125, 241
Analyte 2000, 93
ANL, *see* Argonne National Laboratory
Anthrax, 9, 10, 21, 69, 72, 79, 83, 89, 107, 131, 132, 133-136, 147, 162, 191, 192, 244
Antibiotics, *see* Drugs, antibiotics
Antibodies, 9, 134-135, 191
   assay probes, 81-82, 84
   monoclonal antibodies, 10, 61, 62, 81, 82, 119, 122, 144, 149, 152, 157, 158, 162, 192, 242
   see also Immunoassays; Vaccines
Antibody fragments, 10, 85-86, 93, 149, 152, 154, 162, 192
Anticholinesterase compounds, 5, 44, 60-61, 118

265

M

Masks, filtering, 4, 35-41 (passim), 98, 102, 116, 147, 222
Mass spectrometry, 54, 63, 85-86, 94, 229, 231, 234, 238, 256
Medical personnel, 5, 6
  computer tools, 174-175
  covert exposure recognition, 6-7, 66, 67, 68
  decontamination facilities/procedures, 5, 6, 8, 100-103, 106
  detection and measurement, 43, 59
  intelligence, information, 3-4, 21, 24, 28, 29-33
  law enforcement, communication with, 30-31
  military, 26
  personal protective equipment, 4, 5, 19, 34, 35-36, 41, 42, 100, 103, 106, 108, 147
  psychological effects of attack, 168, 170-172, 173
  response scenarios, 23, 25
  veterinary, 72, 73, 111, 116
  see also Emergency response personnel; Hospitals; Public health officials; Triage procedures
Memantine, 120
Membrane technology, 38, 40
Mental illness, *see* Psychological effects
Methanesulfonate salt of pralidoxime, 118
Methemoglobin, 21, 127, 128, 163, 192
Metropolitan Medical Strike Teams, 25, 28, 55-57, 59, 76, 100, 105, 114, 168, 188
Midazolam, 117
Mine Safety and Health Administration, 35
Miniature technology
  biological agent detectors, 93, 94, 178, 234
  chemical agent detectors, 6, 64, 95, 178, 190, 226, 229, 230, 231, 255
MMST, *see* Metropolitan Medical Strike Teams
Monitoring systems, *see* Detection and measurement of agents
Monoclonal antibodies, 10, 61, 62, 81, 82, 119, 122, 144, 149, 151, 152, 157, 158, 162, 192, 242
Monoisonitrosoacetone, 118
Mustard agents, *see* Sulfur mustard

Myasthenia gravis, 115
Mycotoxins, 11, 21, 76, 154, 156, 157-158, 193, 247-248, 258

N

N-acetylcysteine, 124, 127, 192, 242
NAME, *see* Nitroarginine methylester
National Aeronautics and Space Administration, 38, 84, 255, 256
National Atmospheric Release and Advisory Center, 180-181
National Center for Infectious Diseases, 73
National Coordinator for Security, Infrastructure Protection, and Counter-Terrorism, 18
National Counterterrorism Plan, 25
National Disaster Medical System, 11, 25-26, 28, 167, 188
National Fire Protection Association, 35
National Infrastructure Protection Center, 18
National Institute for Occupational Safety and Health, 34-35, 37
National Institute of Allergy and Infectious Diseases, 148
National Institutes of Health, 140-141, 147-148, 153
National Notifiable Disease Surveillance System, 7, 66-67
National Oceanic and Atmospheric Administration, 180
Nausea, 138, 152
  see also Vomiting
Naval Medical Research Institute, 26, 87, 93
Naval Research Laboratory, 91
NDMS, *see* National Disaster Medical System
Nerve agents, general
  committee's list of, 2, 21, 261
  decontamination, 102, 113, 114
  detection and measurement, 5, 44, 47-51, 53, 55, 56, 58, 60-62, 226-231
  drug treatment, 9, 10, 113-122, 162, 191, 192, 241-242, 259
  personal protective equipment, 4, 36, 37
  psychological *vs* neurological impacts, 169
  symptoms, 60-62, 113; *see also* Seizures and anticonvulsants
  see also Central nervous system; G nerve agents; *specific agents*

Q

R